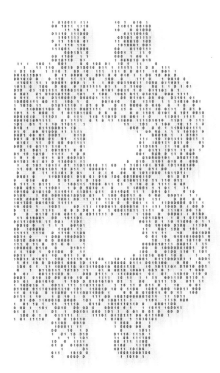

A brief introduction to Bitcoin
Educational, valuable and deep insights

To my mother, Violeta, who is objectively
one of the nicest people in the world.

Table of Contents

1. Introduction 9
2. **What is Bitcoin?** 14
3. **Money and Currencies** 17
   3.1. Definition and functions of money 17
          The basic functions of money 19
          The properties of money 22
          Where does money come from? 23
   3.2. The economic principles behind Bitcoin 26
          The Business Cycle Theory 26
          The Monetary Theory 28
4. **How Does Bitcoin Work?** 30
   4.1. Storing Bitcoin 30
          Asymmetric encryption 31
          Addresses, Wallets and Clients 36
   4.2. Transacting Bitcoin 41
          Transactions 41
          Blocks and the Blockchain 44
   4.3. Creating Bitcoin 47
          Mining 47
5. **Technological Aspects** 50
   5.1. The problem of the Byzantine Generals 50
   5.2. The double-spending problem with fast payments ... 54
   5.3. Decentralization of computing capacity 59
   5.4. Transparency and anonymity 63

5.5.	Technological critique	65
	Trust in hash functions	65
	Dust	66
	Power consumption	67
	The incentive problem	69
	Scaling and the size of the blockchain	70
5.6.	Second-layer solutions	72
	The Lightning Network	73
6.	**Economic Aspects**	**78**
6.1.	Inflation and Deflation	78
6.2.	Bitcoin's planned deflation	80
	Keynesianism and Neoclassicism	81
	The Austrian school of economic thought	83
	Other considerations on deflation	84
6.3.	Scarcity and Stock-To-Flow Ratio	86
6.4.	Liquidity and Volatility	89
	Liquidity	89
	Volatility	90
6.5.	Cryptocurrency exchanges	93
	Distributed denial of service	94
	Failure of payment by service providers	95
6.6.	Further economic considerations related to Bitcoin	96
	Mises' Regression Theorem	96
	Consensus on the rules	99
	Labour Theory of Value	100

  No retraction of transactions 101

7. Suitability as Money 102
7.1. Economical suitability as money 102
7.2. Technological suitability as money 104
7.3. Political suitability as money 106

8. The development of the Bitcoin Economy 108
8.1. Key events and price development 108
  2008 – 2010 108
  2011 - 2012 110
  2013 - 2014 113
  2015 - 2016 114
  2017 - 2018 116
8.2. Adoption and use cases 117
  Crisis currency 117
  Industry sectors 119
  Financial service providers 119
  Donations 119
  Alternative uses of a blockchain 120

9. Closing Words 122
Acknowledgements 125
References 126

1. Introduction

The history of global currencies shows the tendency of national governments to influence their money for economic and political gain.

That's probably a lot to chew on as an introductory sentence. It whispers "conspiracy" and threatens to be a hard subject – money, governments, politics, economics etc. All those not-fun things you didn't sign up for. So, before we get started, I want to assure you that you're stepping onto a fun and educational ride.

I'm guessing you got this book because Bitcoin, the subject that people keep excitedly talking about every few years, is still a bit mysterious – there are still many questions open, after all these years. What is it, how does it work, why does anyone need it, what makes it different enough that it matters, does it even matter, why does it keep popping up in my news feed, what does it change in the world, what can it change for me?

I wrote this book when I, too, was confronted with these questions. The subject Bitcoin caught my attention during my student years, sometime in 2012, and has since constantly fired up my interest in topics I didn't think I could find interesting – investments and banks, which younger me pictured as a boring, serious middle-aged man with a tie and no life. Or cryptography, the quiet kid living in his parent's basement in front of the computer. But, as with all quick and frail judgements, time's wisdom smooths the wrinkles. A master's thesis, a few years of constant learning and a very much alive year-long-interest later, I came to the conclusion that somehow, Bitcoin has managed to turn these topics into the smart, fun and friendly new kid everyone should like, and I want to tell you why. I want to explain

Bitcoin to you, as simply and as deeply as possible. And we will begin right now.

The first thing we need to understand, so that we can truly understand Bitcoin, is where it comes from. What caused it. What burning need the world had which led to the ideas behind Bitcoin, to its creation and ultimately to a mostly warm global embrace, despite regular obituaries and scandals: That need is the need for change of the perceived short-comings of the financial system. This, of course, is an immensely broad topic, so to get an intuition we'll look at some of these short-comings. This is why we'll start with the ability of governments to influence money and we'll work our way from there.

As I was saying, historically, governments tend to influence their money more often than not. Ultimately, the result of governments influencing currency has often been the abandonment of the respective currency due to high inflation rates – the more money on the market, the lower its value. It's easy to forget that this has led most countries to replace their currencies during the 20th century. And without going into too much detail straight away, we'll take a look at a few noteworthy cases which show the potential devastating impact of monetary policy.

First off is the original poster-child of failed currencies, the Weimar Republic in the aftermath of WWI. Upon winning the war, the Allies demanded Germany pay war reparations as part of the Treaty of Versailles. When Germany couldn't or wouldn't, the Allies took over parts of its industrial regions, which forced the German government to cover the war debt and salaries in other ways – by printing money. This, in turn, caused hyperinflation to set in. The value of the currency plummeted and as a result the prices exploded. When Germany introduced a new currency to replace the weak Reichsmark, the exchange rate was one to one trillion! Not to mention the societal and political unrest that followed as a result, and Germany's well-known swerve towards the far-right on the political spectrum.

Another well-known case of hyperinflation is Zimbabwe, whose rampant money printing practices have spurred countless not-so-funny jokes online, mostly about people becoming trillionaires overnight – but to their dismay, without actually getting rich. After Zimbabwe became independent in 1980, its dollar was worth 25% more than the U.S. dollar. An attempted military coup and strong economic decline later, distrust in the financial system and instability followed. To give the impression that everything is going great, the government raised its spending and put in place controls for wages and prices. This way it caused a huge budget deficit, which it attempted to iron over using the money-printing presses. As one can easily imagine, unrestrained inflation followed. It reached 600% in 2004 and 1.700% in 2006. By 2007, inflation jumped to 11.000% and banknotes were denominated in 100 million-dollar-increments. Then came the 500-million-dollar bill which was worth $2,50. Paper money was being used as toilet paper at this point. When the money was finally replaced by a new dollar in 2008, it was equal to 10 billion of the old dollars.

Our last example is Argentina. Its record economic growth came to a sudden halt due to the OPEC oil embargo in 1973. The following political and civil unrest combined with trade and budget deficits foreshadowed an impending recession. But instead of reducing spending or borrowing temporarily to bridge the shortage, the Argentinian government turned to, you guessed it, printing money. The money supply accelerated its expansion following a coup d'état a few years later and by 1982, the Argentinian GDP was collapsing at a rate of 12% per year, an economic downturn previously seen only during the Great Depression. The banknote denominations continued adding zeroes while inflation ran amok, until the currency was finally stabilized by establishing a new peso worth 100 billion of the inflated pesos.

There are many more examples of such practices and their economic cost, and even those countries that survived the century without a currency swap have experienced high inflation rates in the past. A nation's currency is not exempt from the 101 of economic theory - the law of supply and demand. The more money is created, the less it's worth. Printing money is often postulated as a viable option in an emergency, but it's nearly impossible to reverse its effects once the emergency has dwindled. History shows that rampant inflation and recession are the usual results and they require painful steps to be taken to reverse the economic damage.

Today, perhaps more than ever, there are cases of currency manipulation or influencing factors that can impact or even devaluate a currency. And this isn't limited to money printing, either. For example, in October 2012, Mexico banned high-value cash transactions to force more traceable transactions through the digital banking system. At the same time, Argentina started logging and tracking credit card transactions – only four years after the aforementioned economic struggles caused by interference in the monetary policy. Iran's dealings with its own currency, the Rial, are also worth mentioning. The country uses several exchange rates at the same time to circumvent sanctions on the Indian rice markets and thus to buy at discounted rates - to the disadvantage of Indian exporters.

Such interventions can lead to the destabilization and ultimately the collapse of a currency. Financial assets lose their value and the inflation that precedes the crisis causes huge economic costs, which must be compensated afterwards.

To circumvent these and other disadvantages of the traditional, so called fiat-currencies, there are many developments taking place in the field of digital currencies, and more specifically crypto currencies. Here, an attempt is being made to establish new methods to store and transfer value, which due to their immateriality and decentralized nature, have advantages over traditional ones. The crypto currency that has sparked a global phenomenon after stepping into the limelight with its exponential rise in value is, of course, Bitcoin (BTC). Bitcoin is regulated through a so-called peer-to-peer (P2P) network, meaning a direct transfer of information between individuals. This way, Bitcoin enables global value transfer without a centralized controlling entity. Before we move on to learn more about Bitcoin's history and inner workings, here's a quote by its creator, detailing the motivation behind the creation of a decentralized financial system:

"The core problem of conventional currencies is the amount of trust that is necessary for them to work. The central bank must be trusted not to devalue the currency, but the history of fiat money is full of betrayal of that trust. Banks must be trusted to keep our money and transfer it electronically, but they lend it in waves of credit bubbles with a small fraction of security. We must trust our privacy to the banks, trust that they will not give identity thieves the opportunity to empty our accounts. Their massive additional costs make micropayments impossible. [...] With an electronic currency based on a cryptographic proof, without the need to trust middlemen, money can be transferred effortlessly and safely."

- *Satoshi Nakamoto*

2. What is Bitcoin?

Bitcoin is one of the first practical implementations of a concept called *crypto currency*, as in cryptographic currency, which was first described by an engineer named Wei Dai in 1998. The writing of the white-paper detailing the Bitcoin protocol and Bitcoin's first implementation - the creation of its initial code - took place between 2007 and 2009 by Satoshi Nakamoto, a pseudonym whose true owner has not been identified so far and might never be unveiled, despite many individuals laying proof-less claim to the title.

In the first sentence, you probably noticed how I said, "one of the first", not "the first" implementation of crypto currency, and now you're wondering what I meant by that. In the past, there have been multiple other attempts at creating digital money. However, predecessors have failed in one key aspect of Bitcoin's success: its distributed, decentralized nature. Bitcoin is the first cryptocurrency which allows worldwide, direct transactions between two parties via the internet. Please note that in this book, the term "Bitcoin" is used to describe both the payment unit - the currency you transfer - as well as the payment protocol - the system by which it functions and all the processes it entails.

Bitcoins – the units of value transfer – are generated using mathematical calculations. This is done by the distributed computing power of a peer-to-peer network of nodes, or in simpler terms, computers connected directly to each other via the internet. They use software programs called "miners", the first of which, the reference implementation "Bitcoin Core", is an open source project under the MIT license which is driven by a strong developer community. The task of the miners is to validate the authenticity of the Bitcoin transactions made within the network, while earning a reward in the form of transaction fees and newly generated Bitcoin. As a result, the system autonomously verifies and controls itself without using a central entity, such as a central bank.

Before we take a deeper look at money and currencies, allow me to give you a brief intuition on an important concept at the heart of Bitcoin that is often ill-understood - namely, digital scarcity. Sometimes people interacting with 'virtual money' for the first time mistake the terms *virtual* and *digital* for something that is generally more recognizable - files and folders on computers. These are things that you can create out of nowhere, replicate at will and send as copies in unlimited amounts to anyone. So, it's easy to see how this can cause confusion – some might ask what is 'virtual money'? Can anyone simply create it?

In the real world, if I send you a banana, I am no longer in its possession, you are.

In the digital world, if I send you an image of a banana, we then both hold a copy of that image - so I've effortlessly created an additional image of a banana. There are two now.

In the world of cryptocurrencies, however, if I send you a banana-coin, as in the real world, I no longer have neither access to it nor any influence on it. Only you, the new owner, do. The transaction is final and cannot be reversed. The only way for me to get back my banana-coin is if you decide to send it back to me, as an entirely new transaction between the two of us.

Furthermore, it's impossible for me or anyone to effortlessly 'create' new banana-coins. The creation process is very clearly structured, and while it offers everyone the opportunity to participate, it requires a considerable invested effort. Think of it like mining for gold, but instead of muscle labour, this time the work is being done by computers. So let's leave the banana example and return to Bitcoin. This unique property made Bitcoin the first scarce digital object the world had ever seen.

Digital scarcity works by crediting a certain, ever-shrinking number of newly generated Bitcoins to the miners who have completed a complex calculation every few minutes. The calculation ensures the safety of the entire system and we will learn more about it in Chapter 4.3. The built-in scarcity ensures that the number of Bitcoins in existence will never surpass twenty-one million and we will discuss the reasons and the deflationary aspect of this intended limitation in more detail when we look at the economic aspects of Bitcoin in Chapter 6. But for now, we'll attempt to understand money in general.

3. Money and Currencies

3.1. Definition and functions of money

Money is defined as the generally accepted form of exchange and payment in a society that can accept different forms of exchange. Throughout history, this monetary function was fulfilled by a variety of objects, from stones to baseball cards.

Money is usually categorised as

- *Primitive money*, also known as „traditional means of payment" or „primary money", which in addition to its exchange value can have a utility value and can be consumed as a commodity. Good examples are gold and silver. Primitive money is characterized by natural scarcity and can usually be generated and circulated by anyone without a central issuing authority or control body. Think, mining gold.

- *Fiat money*, like the euro and the dollar, which does not require the issuer to convert it into underlying assets like gold or silver. Its acceptance is achieved and ensured by legal regulations. Unlike primary money, fiat money is not suitable for competition - according to economists like Milton Friedman, the introduction of rival, homogeneous currencies would bring its value to zero.[1]

---

[1] (Friedman, 1960)

Nowadays, in most countries the central bank money used for transactions is fiat money, and consists of either

- A currency, i.e. paper money or coins,

- A demand deposit, which usually has no term or has a notice period and is due daily, such as current accounts, or

- A bank deposit, e.g. savings account that is not intended for payment transactions.

Paper money and savings have no value as commodities. A banknote is intrinsically just a piece of paper, and although coins have a certain intrinsic value such as the metal they are made of, this is usually well below the face value. The acceptance of these means of payment in barter transactions is therefore not guaranteed by the intrinsic value of the currency but instead by the trust of the people using it. They trust they will be able to exchange it for any goods or services at a chosen time in the future - a trust usually guaranteed by governments.

As with all goods, the value of money is derived from its scarcity in proportion to its utility. The value of assets, for example, varies depending on the existing amount in relation to the quantity demanded. A pen is relatively cheap, because anyone interested to buy one has the possibility to do so – there are enough pens on the market to satisfy the demand. If, through some inexplicable event, all pens except for one would disappear from the face of the earth tomorrow, some people would suddenly be prepared to pay perhaps a million times a pen's current worth to get their hands on it.

The demand for money depends on several factors, like the volume of trading in a market at any given time, the payment habits of companies or the amount of money saved.

Because the value of money can only be measured by its purchasing power, the value is indirectly proportional to the price level. For example, a constant demand for money would result in price rises because the volume of money increases faster than the offered goods and services. Conversely, prices fall when the money supply cannot compete with the volume of economic products.

The basic functions of money

> *"Money, the classical economists argued, serves three functions: it is a medium of exchange, a unit of account, and a store of value."*
>
> - *Paul Krugman*

A constitutive basic function of money is its suitability as a **medium of exchange**. As simple as it sounds, breaking the simultaneous exchange of goods and services into two separate steps, a buying step and a selling step, represented the beginning of the transition from natural exchange to a monetary economy. The separation facilitates trade because the goods no longer have to be present on the market all at the same time. A trade can therefore be made even between people who don't possess the goods that are in demand. Initially, storable goods such as oils, jewellery, fats, etc. were used as a medium for this purpose. But as trade developed, these were gradually replaced by coins, government and private banknotes and other similar means of exchange.

Another, non-constitutive function of money is its function as a **unit of account**. Equating individual monetary units with a nominal value of one (1) eliminates the need to set exchange rates between any two goods and services. Instead it's enough for the exchange rate to be in relation to the exchange medium, e.g. gold, oil, euro, etc. Here's an example.

The number of exchange relationships between a number of K goods

The figure above shows the exponentially growing number of relationships between an increasing number of traded goods, meaning it gets really complicated, really fast.

For a better understanding, let's consider a market place in a barter economy, where barter goods like cattle and grain are exchanged directly. We can trade a sheep against four sacks of grain. That's one exchange relationship (K2 in the image above). If we now bring in chicken on the market, we have to define two additional prices for the chicken, depending on who we trade with - either half a sack of grain per chicken or eight chickens for a sheep (K3). Bringing then yet another barter good on the market requires us to define three additional prices, one between the new good and each of the existing goods, leading us now to a total of six exchange relationships (K4), and so on. This kind of complexity is the reason why very early on, selected goods were quickly set as the nominal value to which everything else is converted. By introducing a nominal value, this entire process is reduced to a linear complexity. This means a nominal value simplifies defining value and the comparability of goods immensely.

And finally, money is a **store of value**, which is also a non-constitutive function. Storing money allows market participants to transport purchasing power super-regionally and inter-temporally. That's a fancy way of saying that buying and selling can take place in different places and at different times. However, bridging time is not possible if the value of the money changes significantly within a short period of time – something that we call volatility. Modern currencies therefore only fulfil this function if inflation is low and price stability is high.

The properties of money

In addition to the discussed functions, money - regardless of its nature - has the following characteristics:

- *Wide application.* It enjoys a certain degree of acceptance by the population in a socio-economic environment. In many societies, money is first a resource, then due to its own, intrinsic value, begins being used as a medium of exchange.

- *Resistance.* It has longevity and is durable, both physically and in terms of its purchasing power.

- *Good divisibility.* It is easily divisible and is easy to break down into smaller units, so it can be used for smaller transactions.

- *Good consistency.* Despite the ease of divisibility, there is consistency in the value and in the quality of a unit or a fraction of a unit. One gram of gold must have the same value as another gram of gold with the same characteristics.

- *Safety.* It's hard - preferably impossible - to fake, and its authenticity is easily verifiable.

- *Scarcity.* There is a limited amount of it in existence. If we discovered tomorrow that gold exists in abundance and everyone had access to as much of it as they wanted, gold's value would plummet instantly.

- *Transportability and storability.* Finally, money must be easy to handle, store and transport, meaning it needs to be somewhat limited in size and weight.

Where does money come from?

The creation of money is a topic in and of itself, so we'll only briefly talk about the way this works in modern western economies for fiat money.

Normally, a central bank functions as a bank for smaller commercial banks. Its purpose is to ensure price stability by influencing the flow of credit and money on the market. To this purpose, it can create money as it sees fit and can set a short-term interest rate, the so-called "cost of money", for which commercial banks can borrow that money from the central bank.

Money creation in modern economies

In modern economies, the money creation mechanism follows strict controls and regulations and it implies the political independence of the central bank, the mentioned interest rates, a stable inflation rate as the primary objective of the entire process and so on. There are, of course, enough historic examples of states which test the limit of the central bank's independence, or even "print" as much money as they feel necessary, as we have seen in Chapter 1. But this usually degenerates into hyperinflation.

Returning to commercial banks, these can create so-called "domestic" money and bank deposits - this happens every time they issue a new loan. But, contrary to some popular beliefs, commercial banks can only lend as much credit as their assets are worth - deposits from depositors, money from the central bank, equity etc. It's not possible for them to lend more. And usually it's less, since the central bank imposes on them a certain percentage to be held in reserve - the well-known "fractional reserve".

The fractional reserve allows increasing the money supply by using a relatively simple formula. With a 50% reserve, a bank can give out a loan as high as 50% of a deposit. If that loaned money makes its way to another bank, that bank can now again loan out half of it (so 25% of the initial deposit) and so on. This way the money supply doubles, after an asymptotic loan line of 50% + 25% + 12.5% + 6.25% etc.

With a 10% reserve, the money supply increases up to 10-fold following a similar process. With a 1% reserve it increases up to 100-fold and in the complete absence of any reserve it can theoretically increase indefinitely. But the fractional reserve is not related to the way the money is initially created nor does it mean that the banks created it. It's also not tied to fiat money and can be achieved with any kind of money or assets. Increasing the money supply is a direct result of lending. Banning the fractional reserve - by imposing a 100% reserve - would be a de facto ban on lending.

To better understand where all this money comes from, here's an example with bananas. In the image below, we're dealing with a 0% reserve, meaning entire bananas can be lend, and no fraction of a banana needs to be kept as reserve by the banks of Alice, David and Gerome.

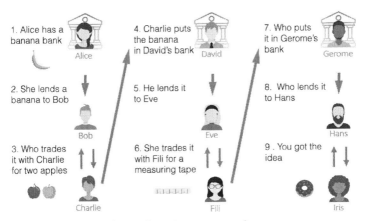

Lending and creating money – or bananas

Alice, David and Gerome's banks have each loaned out an entire banana's worth, although there is only one banana to begin with. This is how out of one banana, we now have three bananas as bank actives, without anyone lending out more than they received.

If you're still confused as to where the additional value comes from, pay attention to the fact that neither Alice, David nor Gerome have any real, physical banana in their bank's possession. What they do have, instead, is a promise worth exactly one banana (plus interest) from Bob, Eve and Hans. These three will have to somehow create a banana's worth of value in the future and repay that promise. So, in a way, you could say that the additional value comes from the future. And someone will have to work for it.

3.2. The economic principles behind Bitcoin

The theoretical roots of Bitcoin can be found in the Austrian school of economics, led by economists such as Ludwig van Mises and Friedrich A. Hayek. These are known for their criticism of the fiat money system and of state regulation through monetary policy, which in their opinion entails tighter business cycles and high inflation. In the second half of the twentieth century, these theories led to a gap between supporters of the Austrian school and supporters of the economist John M. Keynes, the so-called Keynesians, who regard the fiscal and monetary intervention of the state as a necessity for the stabilization of economic cycles.

The Business Cycle Theory

One of the topics that Austrian economists have busied themselves with is the Austrian Business Cycle Theory or Over-Investment Theory. It regards economic cycles as the direct consequence of disproportionately large bank loans. This theory had its beginnings with the Scottish philosopher and economist David Hume in the eighteenth century, as well as the classic English economist of the nineteenth century, David Ricardo. These theoreticians saw that in the eighteenth century, alongside industry, another important institution had been established - banking. Banks had the ability to expand the volume of money by lending, a process in which the economists saw the explanation for the newly created phenomenon of recurrent ups and downs of an economy.

According to the Business Cycle Theory, economic cycles do not arise in a natural money-based economy. In a free market economy, natural money, by its very nature, establishes itself as a means of payment – as we've already seen, gold or silver are good examples. According to the theory, the adjustment between supply and demand of money runs smoothly in these circumstances, so that there are no ups and downs. The excessive expansion of credit and the resulting artificial expansion of the money supply beyond the available reserves are disruptive factors of this optimal balance. The economists criticized the fact that a bank with 100kg of gold could lend up to 500kg of gold (at a reserve rate of 20%) if it pledges to redeem liabilities in cash.

We've already seen how this works, but now it's also important to understand that this process can only work if investors and depositors don't attempt to withdraw all their deposits at the same time - a so-called "bank run". A bank run is an issue since the banks don't hold sufficient funds to honour too many withdrawals. As long as there is no run on the bank, the money lent to the market has the same value as gold, and the bank can add 400kg of gold to the money supply. According to the Austrian economists, this is what leads to a price rise within the currency area in the long term, through artificial inflation. As a result, a mismatch between import and export occurs. Instead of more expensive, local products, cheap goods from abroad are in demand. For foreign companies, however, there is little reason to hold onto the inflationary currency - they exchange it back to gold as soon as possible with the local banks. This results in an increasing gap between reserves and credit volume, which is accompanied by an increasing uncertainty. This can lead to a run on the banks and quickly reverse the country's economic picture – since people will want to get their money out before a catastrophe occurs, which, of course, causes the catastrophe in the first place. The economic term for this is a *self-fulfilling crisis*.

Since today the central bank money is usually fiat money, without underlying assets, there is no redemption obligation in gold or silver to the issuing banks. However, the economic instability remains due to credit, artificial inflation and low interest rates.

The Monetary Theory

Another matter related to the current financial system's shortcomings and which lays at the foundation of Bitcoin is one that was addressed by Friedrich A. Hayek in his work "Choice of Currency". This is called the Monetary Theory. As a means against inflation, Hayek proposes abolishing the monopoly of money creation. Instead of the state or central bank alone, private banks should issue their own, non-interest-bearing certificates that compete against each other on the open market. This way, people could flee undesired, inflationary currencies. This would ensure that only stable currencies prevail in the long run. According to Hayek, it would result in a more efficient monetary system in which only stable currencies coexist.

The concept isn't completely unheard of, as in some countries regional currencies are already used as alternatives or complements to national currencies. In Europe alone there are over twenty regional currencies, like the UK's Bristol Pound or Germany's Urstromtaler, which aim to encourage spending on local products and businesses. Regional currencies are usually backed by the national currency, but this isn't always the case. The Russian Kolion, for example, was a local currency pegged to the value of potatoes and eggs and was meant to support farmers. Sadly, its inventor was arrested, and the currency declared illegal.

Many who share the views presented in the above theories believe Bitcoin is a solution to such shortcomings of the monetary system. Some even go so far as to claim Bitcoin could one day replace banks or offer an alternative. But Bitcoin is not without its own kinks. So, before we get to see how and why Bitcoin addresses such economic aspects and what advantages and disadvantages it offers in later chapters, let's dive into its technological inner workings and attempt to understand how and why it works. Bitcoin is the first scarce digital object, and was only made possible through some very interesting, innovative and perhaps even revolutionary mechanics.

4. How Does Bitcoin Work?

4.1. Storing Bitcoin

To see how the system behind Bitcoin works on a deeper technological level, we'll take a look a few cryptographic methods which are employed in it and in most other existing cryptocurrencies.

Imagine Bitcoin as a ledger in which each transaction between two people is written down as a publicly accessible, immutable list of entries called a blockchain (which we will talk about in Subchapter 4.2). The blockchain is a unique, openly accessible and globally accepted 'truth' about the status of possession of any amount of Bitcoin in existence. Meaning, you don't hold any amount of Bitcoin on your computer, or your phone, or in an actual wallet, as one would believe when thinking of traditional payment systems. Instead, the amount of Bitcoin you possess is recorded on the decentralized blockchain held by millions of nodes.

Now, I said we don't keep any Bitcoin inside a wallet. However, there is indeed a software component in the crypto ecosystem called a "wallet", a term which might cause some confusion. And while this wallet is not used to store Bitcoin, it is used to store the one thing required to access and transact your Bitcoin amounts which are stored on the blockchain: the so-called "private key". A private key is, very simply put, a long string which is used to prove ownership and has its origins in asymmetric encryption. The following subchapters can get a bit technical, but they teach a useful thing or two about computer security, and of course about Bitcoin, so I recommend you stick with me.

Asymmetric encryption

The concept of asymmetric encryption, or public-key encryption, was proposed in 1976 by Whitfield Diffie and Martin Hellman and has become the de-facto standard in the way communication works across the internet today. Every time you hear about SSL or HTTPS, or that you are using a secure connection when surfing online, you are using some kind of asymmetric encryption in one way or another. And while many crypto algorithms are de facto unbreakable, the significant innovation of public-key encryption lies in its drastic simplification of the key management.

For two thousand years, ever since Julius Caesar sent very simple, encrypted messages to his generals, all encryption mechanisms have been symmetric, meaning that both sender and receiver have the same key to encrypt and decrypt a message.

A simple and well-known example for symmetric algorithms is known as the Caesar Cipher. Caesar would encrypt a message by shifting all alphabet letters left or right. For example, he replaced every A with D, every B with E and so on. The key used to encrypt messages was identical to the key used to decrypt them: "3". As in, "shift by 3 letters". Not only did the sender and receiver both have the same key, but they needed to exchange the key in person before being able to encrypt messages and communicate securely – which is fairly impractical for long-distance communication. Another big disadvantage of symmetric encryption is that both sender and receiver need to manage a separate key for each communication partner. As you can imagine, this is an impractical solution that requires huge amounts of effort. So much so, that in the 1980s, U.S. Navy ships stored so many keys to communicate with other vessels that the paper records had to be loaded aboard using forklifts.

To solve many of these problems, one would need to encipher a message using one key, and to decipher it using another, completely different key. With such an asymmetric encryption, Caesar could send secret messages to his general Marc Antony without exchanging any keys beforehand. In fact, he could send him safe messages without having ever personally communicated with him before.

If you're thinking "that sounds impossible", then you're right up there with the biggest and smartest cryptographers of the world up to 1976. Congratulations! But luckily, fast forward a few years, and asymmetric encryption is possible and practical. And while it's complicated in detail, it's simple in outline.

Let's take a look at an example. Caesar wants to send a message to Marc Antony. The recipient of the message, Marc Antony, has two keys. A *public key*, which he can make publicly available to everyone. He can post it on online forums, send it per email to Caesar, or, if email is not an option - this being 50 B.C. and all - perhaps send a messenger on horseback. He could even nail the key to the senate door. The public key can be used by anyone, in this example by Caesar, to encrypt a message before sending it to Marc Antony.

Marc Antony also owns a matching *private key*, which is known to him – only to him, and nobody else, and cannot be deduced from the public key. He can use this private key to decrypt Caesar's message and read it.

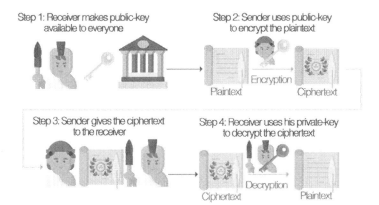

Asymmetric encryption of a message from Caesar to Marc Antony

This way, any communication between sender and receiver only requires the exchange of public keys. Private keys are never transferred.

Such asymmetric encryption is possible thanks to so-called *one-way functions with a trap door*. These make it impossible to deduce the private key from the public key – and thus allow the message to be decrypted only with the help of the private key.

One of the best-known examples of asymmetric encryption is the RSA algorithm published in 1978. The one-way function of RSA relies on the mathematical task of factoring. Factoring a number means identifying the prime numbers which you multiply to produce that number. For example:

64.413 can be factored into

3 x 3 x 17 x 421

3, 17 and *421* are all primes

Interestingly, factoring numbers - a seemingly simple part of primary school education - is seen as an especially difficult task by mathematicians.

ONE-WAY FUNCTIONS

A one-way function is a function which is relatively "easy" to compute complexity-theory wise, but whose inverse function is "difficult" to compute. An example of this is the modulo function, which calculates the remainder of a division. As a simple example, let's calculate the following modulo:

16 mod 13 = 3

That was easy. However, you now cannot use an inverse function to deduce $x = 16$ from 13 and 3 alone, since x can assume different values. To put this differently, even by knowing that the remainder of the division of some number x by 13 is 3, we cannot know the value of x. It could be 16, or 29, or 42 or any one of an infinite number of possibilities.

A special kind of one-way functions are **trap door functions**. For these, even their inverse function is easy to calculate, with the help of additional knowledge - so-called trap door information.

A good metaphor for a trap door function is the operation of a mailbox where anyone can drop a letter, whereas retrieval without the appropriate mailbox key is very difficult. This type of one-way function is used for the asymmetric encryption of messages. The trap door information serves as the key for decrypting the message.

Despite the endeavours of such bright minds as Fibonacci and Gauss, no one has yet discovered a usable and consistent way to factor big numbers. For huge numbers, factoring is an unfathomably time-consuming process, even with super-computers.

Bitcoin uses another algorithm, called elliptic curve cryptosystem or ECDSA, as its one-way function to sign transactions. If this sounds foreign, it's because we haven't talked about signing transactions yet, as we will first take a look at hash functions and digital signatures. Let me just note, for the sake of completeness, that the one-way function of ECDSA is based on the mathematical problem of discrete logarithms in elliptic curves. The search for discrete logarithms of random elements of an elliptic curve is harder than the prime factorization.

Addresses, Wallets and Clients

We now know that we require a private and a public key to secure messages, so let's see how these play into Bitcoin.

A requirement for Bitcoin transactions is that the exchange parties each have an ECDSA key pair. The owner of the private key has access to the funds in the blockchain which are allocated to a corresponding public key. The public key thus serves as a destination for transactions, or as an *address*.

Such addresses are created by generating hashes from the public key (see the sidebar) - for example by using algorithms like SHA-256 and RIPEMD-160 - and then shortening them to 20 bytes.

The result is a so-called **Bitcoin address** and it's a string of 27 to 34 alphanumeric characters that starts either with a 1 or a 3.

HASH FUNCTIONS

Hash functions are one-way functions without a trap door. They are used in cryptography in many ways. Two areas of application are pseudo-random generators and the guarantee of integrity and authenticity in the transmission of a message.

Cryptographic hash functions take a collection of data of arbitrary size and transform it into an alphanumeric string of a fixed length. For example, the hash function SHA-256 used for Bitcoin generates a 32-byte string. This is called a hash value.

The table below shows the generated SHA-256 hash values of similar strings. You can see that even the smallest changes to the original data fundamentally change the resulting hash value. Moreover, it is impossible to recreate the original string from the hash value.

String	SHA-256 hash value
Bitcoin	6b88c087247aa2f07ee1c5956b8e1a9f4c7f892a70e324f1bb3d161e05ca107b
bet coin	a8ebb60a8f60569fb5172a766df5fbcf740003f9eb6972068c093c66557f218d
Bitcoins	b1e84e5753592ece4010051fab177773d917b0e788f7d25c74c5e0fc63903aa9

Thanks to these properties, hash values are often used to securely store passwords in databases. Unlike plain text passwords, hashes are secure even after a third party accesses the database. However, they can also be used to quickly verify the validity of the entered password.

For one private key, any number of Bitcoin addresses can be generated.

Each transaction within the network points to a unique Bitcoin address, or put differently, each address can be assigned a certain Bitcoin amount.

The image below shows an example of a Bitcoin address and its corresponding QR-code.

32ocWb4TCgp2rEAxaBEdkhfiHhTR1aqkAx

Since handling cryptographic keys is not an easy feat for most people, and definitely not a pleasant one for anyone, the storage of the private key is facilitated by the previously mentioned **wallets**. A wallet is the physical repository of the private key. The first Bitcoin wallet was part of a software application written by Satoshi Nakamoto, which is why it is called a "software wallet". Other existing private key storage options include:

- *Mobile wallets* in the form of apps for smartphones and tablets. The private keys are stored locally on the mobile device.

- *Paper wallets* in the form of smart cards, physical coins or bills that secure the private key behind a security hologram. The private key can only be read by damaging the paper wallet. Deposits are still possible with this type of wallet. Disbursements, on the other hand, are only possible if the paper wallet object itself is transferred physically, or if the private key is read, whereby the physical object is damaged and loses its validity as a value carrier.

- *Brain wallets,* which are private keys that can be generated from a long phrase or sentence and need to be learned and remembered by the owner - thus not requiring to be stored digitally. Different service providers offer the possibility to generate such keys online, but many consider them less safe than other storage methods.

- *Hardware wallets,* which are usually tamper-proof devices. They often store private keys in a protected area of a microcontroller from which they cannot be transferred out of the device in plaintext. Hardware wallets can be used securely and interactively as they are immune to computer viruses that steal from software wallets.

Software wallets are usually integrated into larger software **clients**. These are applications which, in addition to key storage, perform several management tasks around Bitcoin, such as mining (which we will discuss in Chapter 4.3). An example of a Bitcoin client is "Bitcoin Core", which was initially implemented based on Satoshi Nakamoto's reference code. Below is a list of the minimum functionalities that a client is typically responsible for:

- *Discovery of nodes.* The client uses a number of methods to locate other Bitcoin nodes, i.e. participants connected to the network.

- *Node connection.* The client initiates, maintains and closes connections to other nodes.

- *Sockets and messages.* The client receives and processes messages from other nodes and sends messages to other nodes via socket connections.

- *Exchange of blocks.* Nodes advertise their blocks amongst each other and exchange them to expand the blockchain.

- *Exchange of transactions.* Nodes pass transactions on and exchange them with each other. The client links transactions to Bitcoin addresses in the local wallet.

- *Wallet services.* The client creates transactions using the local wallet. The client provides administration services for the local wallet.

- *RPC interface.* The client provides an interface for developers with various operational features to manage the local wallet.

Most clients supplement this functionality with a user interface.

To be able to perform some of the above-mentioned functions, a client might need to download the entire blockchain and preserve it locally. However, due to the size of the blockchain which surpasses multiple gigabytes, some users prefer to use thin clients like "Electrum" or "Multibit" when mining. These hold and manage the blockchain on a remote server and eliminate the need to download it. This comes with the downside that users need to trust the providers of the thin-clients and trust that they are using the most recent blockchain. This is criticized by many as bad practice, since relying on others to hold the blockchain weakens the decentralized nature of Bitcoin.

4.2. Transacting Bitcoin

In the peer-to-peer network which comprises the Bitcoin ecosystem, two types of data are transmitted between all participating nodes. These are transactions and blocks of transactions.

Transactions

Transactions are cryptographically signed operations (see sidebar) involving the transfer of Bitcoin ownership from a collection of one or more Bitcoin addresses to another collection of one or more Bitcoin addresses.

Bitcoin transactions work differently than fiat money transactions. They are redirecting Bitcoin amounts from previous transactions to new addresses, rather than just changing the current amount in an account. All Bitcoin

DIGITAL SIGNATURES

Digital signatures are a combination of asymmetric encryption with hash functions. The hash function of an encrypted message is encrypted with the private key of the sender. The resulting digital signature is sent together with the encrypted message, allowing each recipient to decrypt the signature using the sender's public key. This way, anyone can verify if the message matches the original and if the identity of the sender is correct.

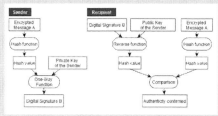

Creation and verification of digital signatures

The illustration above shows the steps involved in digital signing and the verification of the signature.

1. Generation of the digital signature
   a. An encrypted message is input into a hash function. This returns a unique hash value.
   b. The generated hash value and the sender's private key serve as input for a trap door function. The trap door function returns a clearly identifiable digital signature.

2. Verification of the authenticity of the message
   a. The digital signature is decrypted with the sender's public key. A successful decryption proves that the publisher of the public key is the sender of this message. The result is the hash value of the encrypted message.
   b. The resulting hash value is compared to the hash value of the encrypted message. If both hash values match, then the message has not been modified.

transactions are thus an ongoing, connected chain of money movements from owner to owner.

In the image below, you can see how a Bitcoin transaction is constructed and how it's tied to other transactions. Transactions include inputs and outputs, each of which can reference multiple Bitcoin addresses.

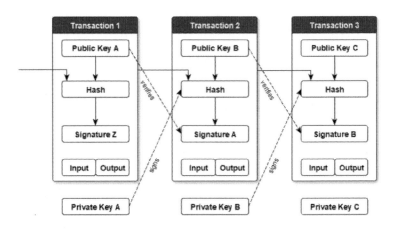

The structure and references in a Bitcoin transaction

Inputs are references to amounts from previous transactions. Each input includes a digital signature confirming authenticity and unlocking the amounts from its previous transaction. The signature is generated with the private key matching the input address so that the transaction can only come from the owner of the amounts.

Outputs reference the future owners of the transacted amounts. They contain the Bitcoin addresses to which the Bitcoin amounts will be credited to. They can also entail rules and criteria regarding the change of ownership. A typical output references a single Bitcoin address, i.e. it requires a single ECDSA signature. However, the references stored with the inputs and outputs are not individual addresses, but scripts. The scripting system used by Bitcoin allows for complex operations and criteria. For example, it is possible for an output to require multiple signatures, two ECDSA signatures, or even a 2 by 3 signature scheme. Due dates for executing the transaction are also conceivable. The output script specifies the requirements that must be met to unlock the transferred amounts. If the transaction is later redirected to an input, the input must fulfil the requirements defined by the output script.

In a transaction, the sum of all inputs must be equal to or greater than the sum of all outputs. If the sum of the inputs is greater, the difference is treated as a transaction fee and credited to the miner verifying that transaction. These fees

DIFFICULTY

The verification of a block lasts on average 10 minutes, regardless of the total computing power of the network. This tends to happen faster if there is a high amount of new, incoming computing power. To keep the verification duration stable, a degree of calculation complexity, a so-called "difficulty", is determined. This value, which denotes the required number of leading zeroes in the next hash, is dynamically adjusted every 2016 blocks and depends on the duration used to solve the previous 2016 blocks.

For example, if the generation of the previous set of blocks took less than two weeks, the difficulty increases, by lowering the threshold under which the generated hash values must lie. The more zeros precede the hash value, meaning the smaller the hash value, the more difficult is its generation. Each distributed block that does not reach the required difficulty is ignored by the network.

prevent excessive numbers of very small transactions and ensure the long-term retention of computing power. A typical transaction fee is measured in Satoshis per byte - a **Satoshi** being the smallest unit of a Bitcoin, which represents one hundred millionth of a Bitcoin. Since transactions take space in a block, bigger and more complex transactions can incur higher fees. A fee can also be increased voluntarily to speed up the verification of the transaction.

Blocks and the Blockchain

After a transaction has been created, it is broadcast to the entire Bitcoin network and is kept by each receiving node in a temporary "waiting area" called a memory pool, or **mempool**. Then, during the validation process, each node decides for itself which transactions it adds to a so-called block. The decision is usually based on the fees entailed in the transactions, which will be allocated – or "rewarded" - to the mining node.

The validation process ensures that the transaction structure is correct, and that the transaction is not included in any previous block. Nodes can also choose to reject transactions with small or no fees. This can cause fee-less transactions to take longer until they are validated.

The node then attempts to compute a hash value from the assembled block. The hash value must meet certain criteria which we will look at in the next paragraph. If the node succeeds in computing that hash value, that block is distributed to all participants within the network. The receiving nodes recheck all contained transactions and then attach the block to their held blockchain. The blocks are thus ordered sequentially into a ledger and the order and consistency of the blocks in the blockchain is ensured through cryptography. In order to attach another block to the existing blockchain, the calculating node must use the longest chain of blocks known to it when calculating the new hash, since it must refer to the last block in its calculation.

As we've already noted, each hash value is calculated based on the hash value of the previous block. Additionally, it's also based on the transactions chosen to be part of the current block as well as a randomly generated number called a **nonce**.

The computing node generates and checks a huge amount of different hash values from the same data, by setting a different nonce each time. It does this until a hash matching all requirements, including the **difficulty** requirement, is found.

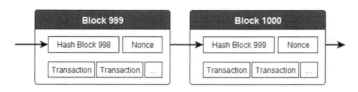

Components of a transaction block

Because the longest chain is used as the basis for new blocks, the blockchain may branch out if participants compute hashes based on different blocks. This is why, in practice, a transaction is usually only considered irrevocably confirmed after it's already six blocks deep in the blockchain - meaning it has been verified and included in blocks for approximately an hour. That's six times the average 10 minutes it takes to calculate a valid hash and thus solve a block.

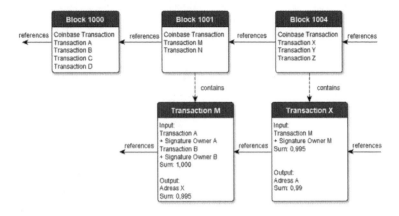

The relationships between transactions, blocks and the blockchain

As we've seen earlier, each node participating in this calculation - or **mining** - process usually holds a copy of the entire transaction history. It's intuitive but perhaps interesting to note that this chain of blocks, which is held in a hash tree structure, can be traced back to the first ever generated block, the so-called **genesis block**.

4.3. Creating Bitcoin

Mining

Mining is the term describing the validation calculation process we've discussed in the previous subchapter and all the surrounding infrastructure and the ecosystem related to it. It involves a reward associated with newly generated – or **mined** – Bitcoins which is rewarded to whoever manages to identify the latest valid hash and thus validate a block. Every user who has internet access and an installed Bitcoin client can participate by using her or his computing capacity. However, due to a nearly exponential increase in the computing power required, mining has become a highly specialized activity over the years.

Until 2010 even a conventional CPU could achieve a positive balance between profit and electricity costs. Later GPUs were used for computation, followed by FPGAs, which offered similar computing power as GPUs with lower power consumption. Starting 2013, so-called ASICs were introduced on the market. ASICs are hardware chips that have been designed from the ground up for the sole purpose of Bitcoin mining. These are far superior to earlier technologies, in terms of both the achieved computing power measured in hashes/s and the power consumption. ASICs are a primary reason for the accelerated increase of the computing capacity of the network.

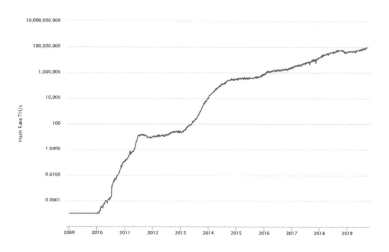

Computing capacity of the Bitcoin Network over time (Hash rate in TH/s)

Due to the high hardware requirements solicited by mining nowadays, mining is increasingly run by specialized companies - so-called mining pools. These combine the computing capacity of many sources, including private individuals. This way miners can rely on a constant stream of income proportional to their invested capacity instead of hoping for an occasional lucky shot, as long as mining is profitable.

A special type of transaction that we haven't talked about until now, the "coinbase" transaction, has no input and exists exactly once per block. The coinbase transaction is generated and added to the block by the verifying node itself. It pays the node operator a *reward*, which comprises of a fixed, predefined number of newly generated Bitcoins.

The reward consisted of 25 BTC for each block until November 2016 and was reduced to 12,5 BTC until May 2020, after which it is further reduced to 6,25 BTC. The reward is halved every 210.000 blocks, a recurring event which has become known as "the halving" and which will, in the long-term, reduce the reward to zero and thus result in a maximum total of twenty-one million Bitcoins. The chart below illustrates the inflation rate, e.g. the amount of newly mined Bitcoins brought into circulation, as a consequence of the halving and the resulting long-term supply growth of Bitcoin.

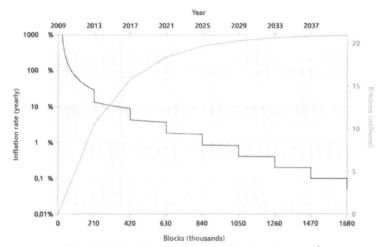

Bitcoin's falling inflation rate and limited supply growth[2]

The coinbase transaction attributes the newly generated Bitcoins as well as all transaction fees contained in that block to the node operator. Like with other transactions, the coinbase transaction can assign the output to either a single address or to multiple Bitcoin addresses. The output is always the sum of the reward plus all transaction fees included in the block.

---

[2] (Bitcoin Inflation, 2015)

5. Technological Aspects

5.1. The problem of the Byzantine Generals

> *"It is not sufficient that everyone knows X. We also need everyone to know that everyone knows X, and that everyone knows that everyone knows that everyone knows X - which, as in the Byzantine Generals' problem, is the classic hard problem of distributed data processing."*
>
> - *James A. Donald*

A fundamental problem in distributed networks is finding consensus in the presence of faulty or defective processes. A dependable system must manage the failure of one or more of its components when it either permanently crashes, repeatedly boots-up and shuts-down, or behaves byzantine, meaning damaging in the worst possible way. For example, nodes may deliberately and maliciously misrepresent or contradict information or disseminate information to make it impossible to unify the network. This problem is described in the literature as the "problem of the Byzantine Generals". It occurs specifically in the case of Bitcoin when the entire network needs to agree on the secure and irreversible validity of transactions, i.e. when it needs to decide which blocks are the generally accepted ones.

The problem of the Byzantine Generals can be described as follows: A number of generals position their armies outside a besieged city they want to conquer, and they need to choose the time of their attack. They know that they can only be victorious if at least half of them attack at the same time. But if they don't coordinate the time of the attack well, they are outnumbered and will lose the fight. They also suspect that some of the generals are disloyal and will send out fake messages regarding the time of the attack. And since the generals can only communicate with each other via messengers on horseback, they have no way of verifying the authenticity of a message. So we ask ourselves, how can consensus regarding the timing of the attack be reached in these circumstances, despite the lack of trust and without a central governance entity.

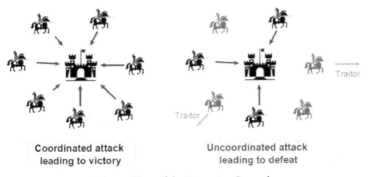

The problem of the Byzantine Generals

Prior to Bitcoin, this problem was considered perhaps impossible to solve. Computer scientists declared in 1982 that the generals' problem can at most be reduced to a "commander and lieutenant" problem, in which all lieutenants must act in accordance with the commander's orders, as long as they are loyal. They have shown that the problem can only have a solution if more than two-thirds of the generals are loyal.[3]

Bitcoin seeks to provide a universal solution to the problem, through which the loyalty of more than 50% of the computing capacity is enough to reach consensus. In this case, disagreement is temporarily accepted and the solution to conflicting opinions is determined by majority vote. In other words, consensus is reached because computing resources are scarce and because one's own performance is used as a vote. This makes it possible to build applications in a decentralized way, which would not have been possible earlier without a central controller. For this reason, the technology behind the Bitcoin protocol - the blockchain and distributed consensus - is increasingly used in other areas too, not only in transaction verification.

Going back to our example, if we regard the generals as an analogy to the individual mining nodes of the Bitcoin network, then the solution offered by Bitcoin looks like this: At any time, any general may announce an attack time. The attack time that is first heard is considered the official plan. However, the transmission of a message is not immediate, which is why different generals can hear different plans first. Therefore, once a general receives a plan, he must solve a difficult mathematical problem based on that plan - a task which preferably lasts multiple minutes on average. As soon as one of them finds a solution, he adds it to the plan, thus generating a new plan. He sends the new plan to the other generals. After that, everyone who hears it begins to expand on this new official plan, which should again take multiple minutes.

---

[3] (Lamport, Shostak, & Pease, 1982)

Every general is working on the most frequently expanded plan known to him, in order to extend the longest living solution that he has heard until then. After a solution has been expanded multiple times, the attack time contained in the longest chain of calculations (e.g. in the longest plan) is considered to be the true attack time, because it necessarily required more than half of the calculation capacity of all the generals to create it. Put differently, the mere existence of this longest block-chain is proof that the majority of generals (over 50%) were involved in its creation. No attacker with less than half the computing power could have produced another chain of similar length during the same time. This scheme is called "proof-of-work" and the resulting hash-tree structure of the blockchain is shown schematically below.

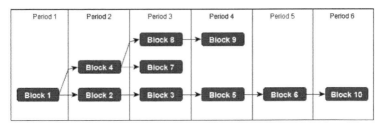

The hash-tree structure of the blockchain

One aspect of Bitcoin complicates the problem described above. In the problem of the Byzantine Generals a *final decision* exists with which all nodes agree. For Bitcoin, this requirement does not exist, so no deterministic solution is provided. The Bitcoin protocol is highly probabilistic and random. It guarantees that the decisions will eventually converge, but without setting a final date. This *self-stabilizing* solution adopted by Bitcoin was described by Dijkstra in

1974 and published in 2006 by Angluin and others, and applied to consensus finding in distributed systems. *Eventually* an agreement on the validity of each block takes place with absolute certainty. At the same time, it cannot be ruled out that at any point in time a longer chain will appear, which supersedes the previous chain. As a result, all transactions in the shorter chain lose their validity from the time the branch out - the so-called "fork" - takes place. Only those transactions remain valid, which have the same inputs and outputs in the long chain as in the short chain. The discrepancy between the two chains may allow, in certain circumstances, the double spending of the same amounts. This is an issue we'll discuss in the next subchapter.

From a technological point of view, such a fork (or branching out) can be expected at certain intervals. A major such fork took place in March 2013. A new version of a Bitcoin software client, due to a lack of backward compatibility, caused older versions of the same client to fail to recognize the generated blocks. This led to old clients working on their own blockchain. The fork existed for 24 blocks - about 6 hours - and disbanded after most of the network switched back to the old version.

5.2. The double-spending problem with fast payments

Bitcoin's proof-of-work scheme prevents double spending for *slow payments* where the payee waits for the transaction to be acknowledged in at least one block. Each new transaction goes through two stages of verification, once after creation and once after it has been included in a block, as described earlier. As a result, a double spending of the same Bitcoins would require a very high computing effort for the generation of fake blocks. This type of double spending is discussed in the next chapter, as part of the larger issue of centralizing computational capacity. In the current chapter we discuss the risk of double spending in fast-paced scenarios.

Unlike conventional banking systems, Bitcoin preserves the entire history of the transactions, which can cause information asymmetries. This means not everyone will know the latest status of the blockchain at the exact same moment. Propagating a payment to the network typically has a delay of less than three seconds. The confirmation of a transaction lasts on average 9 minutes and 54 seconds with a standard deviation of 881.24 sec (approx. 15 min.)

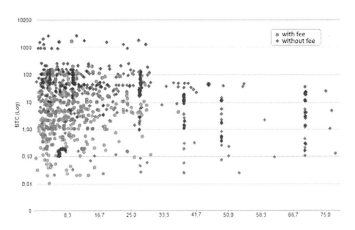

Duration of transaction confirmation (in minutes) based on paid fees[4]

---

[4] (Bitcoin Stats, 2013)

Such short-lived asymmetries may suffice in practice to issue the same Bitcoin amounts multiple times by triggering transactions at different addresses simultaneously. This problem is described in the literature as the *double-spending problem*.

Bitcoin is used in a number of fast-paced scenarios, for example over-the-counter purchases. And while this is slowly being taken over by second-layer solutions - which we will discuss in Chapter 5.6 - sellers right now can't always wait for the complete confirmation of a Bitcoin transaction – or for that matter, not even for a partial confirmation. If you buy a coffee, you expect the process to last seconds, not minutes and certainly not an hour. In the literature this is called a *prevention of activity* for enterprises which are characterized by fast service times. Providers are therefore prompted to conduct and accept low-value transactions without confirmation.

Such so-called "zero-confirmation transactions" allow double-spending attacks in which the exchange of goods is completed before the transaction verification has been conducted. Here's what that would look like:

To perform a successful double-spending attack, attacker A must make the vendor V accept an unconfirmed transaction TR_V. At the same time, A generates another transaction TR_A. This has the same inputs as TR_V, i.e. both transactions use the same Bitcoins, but they reference a different output address. For example, instead of the address of the vendor V, attacker A could reference an address under his own ownership.

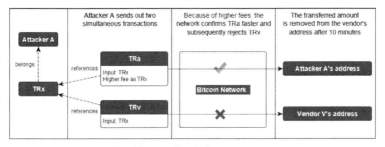

Double spending in fast payments

The attack could be successful if the vendor V receives transaction TR_V, which was meant for him, but the network instead confirms transaction TR_A and the trade is settled before the first block acknowledgment, i.e. before the Bitcoin amount is taken out of the wallet.

If both transactions are broadcasted at the same time, both have a similar probability of being included in the next block, since the network nodes always accept only one transaction with the same inputs. To help his cause, attacker A could modify the output of TR_A to influence the size of the transaction fee. That way he can get a better confirmation probability for TR_A.

This type of attack is considered feasible in practice if no countermeasures are taken. Therefore, extensions exist which are considered ways to protect against double spending. For example, such extensions are "listening periods" in which the network checks all incoming transactions for equal inputs, or "observers" which point out fraud attempts.

Other solutions once in use were so-called "green addresses". With these, a vendor who considered the owner of the address trustworthy could accept zero-confirmation payments. However, this method was highly criticized for requiring third party trust, a necessity which Bitcoin is meant to circumvent. A better way in which the double-spending problem can be mitigated is by introducing a second-layer mechanism for fast payments. We will discuss an example of an existing second-layer solution, the „Lightning Network", in Subchapter 5.6.

5.3. Decentralization of computing capacity

> *„The viability of the system depends on the assumption that nobody will ever gain [more than 50% of the network's computing] power. "*
>
> *- Becker et al.[5]*

Bitcoin is designed as a distributed network of independent miners and assumes that no single entity can hold more than half of its total computing capacity for any significant amount of time. If such a centralization of capacity occurred, it would allow the attacker to determine which transactions are allowed and which are not.

Although the processing power of the Bitcoin network surpassed 80 exahashes/s (80 x 10^{18} calculated hashes per second, or about ten times more than grains of sand on earth) in 2019, most of this power is shared by an oligopoly of mining pools. There are only a few associations of miners who are involved in the collaborative creation of blocks and they share the rewards according to the computing power contributed.

---

[5] (Becker, et al., 2012)

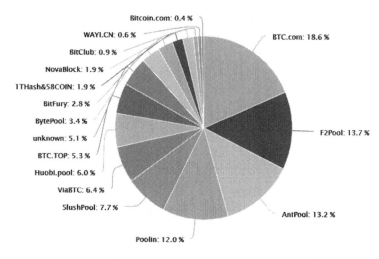

Hashrate distribution amongst mining pools in percent[6]

Even if an aggregation of computing capacity is conceivable, from a financial point of view, there is an incentive for any node to remain honest. If a profit-oriented attacker is able to sustainably combine more processing power than any honest node, he has two alternatives: he can either use the processing power to rewind his own payments and thereby cheat other Bitcoin network participants, damaging the network's credibility, or he can generate new Bitcoins. The second alternative and thus compliance with the rules should be more profitable. This allows the attacker to harvest more Bitcoins than the rest of the network, rather than undermining the system and the validity of his own profits, basically sabotaging his own interests. This

---

[6] (Blockchain Info Pools, 2019)

makes the merger of profit-making entities that sustain more than half of computing capacity unlikely.

A much more likely scenario is an attacker who, unlike mining pools, could only achieve a time-limited takeover of more than 50% of the network rather than a sustained one, with the intention of generating one-time gains rather than harming the network. While digital signatures prevent any manipulation of already existing wallet amounts, an attacker would have the ability to spend his Bitcoins twice, thereby making a profit or preventing the inclusion of transactions in the blockchain, thereby jeopardizing the integrity of the system.

To imagine this attack scenario, we must first make a few assumptions. We assume that both honest miners and the attacker have the same mining expenses to cover, but that the attacker is able to have blocks generated faster within his budget. Furthermore, we assume that we can derive the damage caused only from the transactions affected, without taking other factors into account.

The attacker has a fixed budget B available for calculating a certain number of hashes. This is the expected number of hashes to create a branch of the blockchain of the length of B. The attacker can use the hashes at the same time or split them arbitrarily, i.e. he can temporarily have more computing power than the rest of the network. Under these conditions, the attacker could create a branch of seven blocks in length. Thus, the transactions confirmed by him would be considered valid by most exchanges. Most exchanges await a 6-block confirmation until they make irrevocable decisions, for example exchanging Bitcoins into euros or U.S. dollars.

Putting all of this together, a scenario for a profitable attack target would look like this: The attacker deposits an amount of x U.S. dollars into a Bitcoin exchange, withdraws the amount after 6 blocks, then rewinds the network and spends those x U.S. dollars a second time.

Since the attacker has a finite budget, any attempt to create seven blocks alongside the actual blockchain carries a degree of risk, since there are still other mining parties in the game which hold a non-trivial portion of the capacity. He can therefore expect either a potential profit of x USD or a potential loss of B. Since exchanges usually only allow deposits up to a certain amount, we can be nearly certain that the risk amount B exceeds the profit potential x many times over.

While such types of attacks could be profitable in the long run, a single losing streak on limited budgets will lead to bankruptcy. This is similar to the martingale betting system, in which a gambler at the casino doubles his bet after every losing toss. A gambler with infinite wealth will eventually win back his losses and make a profit, but the exponentially rising bet size combined with the fact that strings of consecutive losses occur more often than intuition would suggest, usually leads to a catastrophic loss. This is called the gambler's ruin.

Due to this high financial risk, it's unlikely that such an attack would be carried out for profit-making purposes. On the other hand, a politically or otherwise motivated attempt to damage the network in the long term, where costs play a secondary role, is conceivable.

5.4. Transparency and anonymity

> *"[Anonymity] is definitely a concern, and it is definitely part of the reason I say that Bitcoin is an experiment."*
>
> *- Gavin Andresen (former Bitcoin developer)*

While the centralization of computing capacity doesn't seem to pose a threat to Bitcoin, there is another, similarly often repeated misunderstanding of the way blockchains work. Some people claim that Bitcoin is fully anonymous - a claim often uttered in the same breath as some statement about illegal drugs and money laundering. However, "An Analysis of Anonymity in the Bitcoin system" by Reid and Harrigan shows us that Bitcoin is de facto not anonymous. And while this is great news for transparency-lovers, it entails privacy disadvantages which we will shortly take a look at here. Transactions may be conducted in a manner that obscures the identity of the parties, but in many cases, it can still be determined.

There are generally big difficulties in maintaining anonymity in networks and online services when partial user information is exposed. Such difficulties arise in different scenarios, for example when attackers combine multiple sources of information to identify users from limited anonymous (so-called "pseudonymous") networks. Or even statistical data matches between different but related data sources. For example, Netflix data can be deanonymized using IMDb data with similar user content.

As long as a Bitcoin address is not clearly assigned to a person or a company, the identities of the transaction participants are anonymous thanks to the decentralized nature of Bitcoin. Then again, the transactions on the IP level are traceable, so complete anonymity of the payment can only be guaranteed by concealing your IP address. And, of course, this all breaks down as soon as the owner of a Bitcoin address can be clearly identified, for example by an exchange or a vendor. Centralized service providers, like exchanges, mixers, or wallet providers have user information that can link user activity and addresses. Active analyses of the blockchain can make payments traceable.

Not only that, but all regulated Bitcoin exchanges are required by law to trade transaction information with the police and other legal entities, following money-laundering regulations in Japan and the United States. They require verified personal data as a prerequisite to participate in the market. Thus, no anonymous transactions can be carried out through exchanges.

And since the transparency of the blockchain enables the tracking of transaction flows, it can easily pose a threat to individual privacy as well as lead to competitive disadvantages for companies. To avoid this, usually a new address is generated for every incoming transaction. This way, the payment sender cannot see any other payments that took place on the receiver's end. But even this small glimpse of anonymity reaches its limits as soon as, for example, one company concatenates all payments of a financial period to exchange them into another currency - and the concatenated address can be tracked back to all purchases and money flows performed by that company.

5.5. Technological critique

Now that we've seen the bigger kind of technological challenges that Bitcoin faces, let's also take a look at some smaller issues which to this day seed long-term discord and some of which might not be easily solved "on-chain" - on the Bitcoin blockchain itself - but rather with second-layer solutions that are applied on top of Bitcoin. We'll talk a bit about the general confidence in hash functions which is expected for the functioning of the network - a topic which many gloss over, then about nearly worthless microtransactions called dust and the expensive power consumption of the network, and finally the incentive of mining nodes to withhold high-fee transactions as well as the ever-growing physical size of the blockchain.

Trust in hash functions

One reason why the use of hash functions had not been widely explored in distributed systems before Bitcoin is that, from a traditional perspective, they are a strong assumption. What this means is that no hash function behaves perfectly, e.g. there are cases in which two distinct pieces of data will have the same hash value. This is called a hash collision. And although Bitcoin is certainly not based on unsustainable properties of hashes, since collision resistance has been proven to a certain extent, any hash-based protocol will have a positive, albeit negligible, probability of failure.

Consider the „random oracle" model - a theoretical, cryptographic model that models ideal hash functions. Consensus protocols such as Bitcoin are demonstrably secure within the random oracle model, e.g. when no collisions can occur. But some studies show that the real-world application of the protocol can be uncertain. What this means is that in very rare cases – so rare, that they probably will not occur in millions of years – validated blocks could have the same hash value.

It is also interesting to note that the first applications of hash-based methods were implemented by Dwork and Naor, two leading figures for distributed consensus building. However, these had not considered using such methods in consensus algorithms, for the reasons mentioned above.

Dust

Dust refers to transactions that transfer very small amounts of Bitcoin. Since Bitcoin stores every single transaction in the blockchain, each transaction requires memory and processing power to be verified. In the past, individuals have used very small transactions of as little as one Satoshi to store information in the blockchain, causing legitimate transactions to stack up and the network to slow down. This was seen as a big potential issue initially but has been more or less solved by setting a minimum size for a transaction.

Dust, however, is still problematic in cases in which the transaction fee surpasses the value of the transaction output – meaning that even with a fixed minimum size, very small transactions can be stuck in miner's transaction backlogs for long or forever as they are unprofitable.

A solution for users is to use wallets that allow consolidating addresses holding dust into one. The caveat here is that, by consolidating addresses, they become identifiable on the blockchain as belonging to the same owner. Meaning, if you are identified as the owner of one of those addresses, you are identified as the owner of all of them.

Nevertheless, you can rest assured that, despite not being completely solved, the topic is a non-issue for most intents and purposes.

Power consumption

Electricity consumption is the favourite critique point of many crypto currency opponents, as postulated by Paul Krugman in his article "Adam Smith hates Bitcoin". Here, he refers to the need to spend real resources on the production of Bitcoins as a drastic step back and underlines this argument by using the criticism of Adam Smith, in which gold and silver are described as a "dead inventory" with no added value for the country.

Although this does not properly illustrate how Bitcoin works, wasting resources is still a valid critique point. Before we discuss it further, note that it's not just the generation of Bitcoins per se that consumes processing power and electricity, as is often wrongfully stated, but instead the process of transaction verification via mining and thus the Bitcoin network's security itself - as we have seen previously. Newly generated Bitcoins are merely a reward for strengthening and improving the reliability of the infrastructure. And since every monetary system requires an infrastructure to function, which will invariably lead to spending resources and energy, Bitcoin is no different in this respect than other systems. A more appropriate question would therefore be: Which monetary system causes a higher waste?

Take into consideration gold, and the amount of time, lives and resources invested into its mining, transportation and safe-keeping throughout history? Not to mention the forced mining camps in

African countries, child-labour and the pollution. Or we could take into consideration the energy consumed to sustain the current financial system, with its approximately 30.000 banks and similar FIs worldwide, its servers, branch offices, ATMs, middlemen etc. Even after excluding non-monetary services offered, banking consumes an estimated 100 terawatts of power annually, which far surpasses Bitcoin's consumption.

Another argument speaking against this critique is the adaptability of the computing power to the Bitcoin value. This suggests that Bitcoin is a rather efficient way to protect the consensus process from attackers by making attacks economically unprofitable. As the Bitcoin value increases, so does the incentive to invest more computing power, yet Bitcoin does not consume more computing capacity than necessary to maintain the integrity of the system. Not more will be invested in mining than would be profitable for the miners. The amount of rewards falling over time also ensures that the use of computing power is limited in the long run and that it will eventually decrease together with the profitability for the miners.

Last but not least, there are conservative studies indicating a clear trend among miners to opt for areas with abundant renewable energy. Bitcoin miners seem to be fairly evenly distributed across the planet and are predominantly focused in technologically advanced, sparsely populated, mountainous regions traversed by rivers, with an estimated 60% of global mining happening in China. This is a result of the rainy season in the hydroelectric-heavy provinces of Southwestern China, which makes electricity highly affordable. And there are cases where electricity is cheaper than free - negatively priced electricity produced by solar and wind energy during periods of oversupply. By investing this energy into mining, electricity networks can be stabilized and simultaneously return a profit. Adding things together, the studies indicate that 74% of Bitcoin mining activity worldwide runs on renewable energy.

Nevertheless, because I believe we should take care of the planet for future generations, perhaps it's not fair to compare Bitcoin to the monetary services of the entire current financial system, or perhaps it's worth looking at more energy efficient alternatives to Proof of Work. And this whether or not the consumption is considered wasteful.

To this end some cryptocurrencies are already experimenting with the so-called "Proof of Stake", a validation scheme in which the creator of the next block is chosen via various combinations of random selection and wealth or age (i.e., the stake). The more coins you own, or the longer you've been part of the network, the higher your chances are to be the creator of the next block. But Proof of Stake, for all its energy-friendly merit, has yet to prove it can keep a network secure and valuable.

The incentive problem

In the long term, perhaps in a hundred years, the falling reward rates for transaction validation will ensure that the profitability of mining can only be guaranteed through transaction fees. This has two possible downsides.

The first is the rather improbable idea that transaction fees will not be sufficient to keep miners interested, and the mining activity will recede considerably, leaving less computational power to secure the blockchain and thus making the blockchain more vulnerable. The true risk of this depends of the equilibrium levels which the fees will reach – and whether governments or other institutions would have enough capacity to make use of such a vulnerability, should they intend to.

If this is not the case, and transaction fees will be sufficiently high to fund the continuation of mining operations, then the second downside is the dependence on transaction fees alone. This provides a financial incentive to review in particular such transactions that offer higher fees. Because of this, as transactions are routed through the nodes, there is an incentive not to communicate the knowledge of high-cost transactions to competing nodes. That way, the competition can be prevented from authorizing the transaction more quickly and withholding the associated fees so that the informed node may be able to claim the fee at some point.

Such a scenario is conceivable if transactions are distributed through a central entity - as in the case of large exchanges. In consultation with a high-performance mining pool or a mining company, the two parties could charge higher fees than the rest of the network for the exchange's participants – of course, with the risk of being called out and staining their reputation in the process.

Luckily, this problem doesn't exist for transactions distributed in a decentralized manner to the entire network - as is the case for individual, independent wallets.

Scaling and the size of the blockchain

Another criticism of Bitcoin is the continuously growing blockchain. Some critics complain about the decreasing manageability for individual nodes, for which it may become unreasonable to preserve the entire blockchain locally.

By some, this problem is considered less critical because the blockchain can be clipped at certain intervals. But to circumvent the forking risks mentioned in an earlier chapter, the clipping would have to occur at a point in time that lays far back, which critics also consider risky. To completely avoid this risk, several proposals have been made over time to implement parallel supplementary chains, to add second-layer solutions to lighten the burden on the blockchain, or to apply different compression techniques to the existing blockchain. Even so, scaling of the network remains an issue, as even bigger blocks could not accommodate a global scale system requiring hundreds or thousands of transactions per second.

The topic of block size has been the cause of tremendous discord in the Bitcoin developer community. It has even led to a fork which created a separate coin out of the main Bitcoin blockchain in 2017, called Bitcoin Cash.

Proponents of Bitcoin Cash argued that the blockchain size is a non-issue, as hardware and infrastructure advancements will enable anyone to run a full-node, e.g. to download the entire blockchain, even in the future. Therefore, they decided to increase the allowed size of single blocks, in order to better scale the network by allowing a higher number of transactions per block. The majority of the Bitcoin community, however, spoke against this measure, arguing that the bigger the blockchain, the less decentralized the network will become long-term as small players can't afford to keep up. This spearheaded the decision to keep the initial block size until an increase becomes absolutely necessary, trusting that second-layer developments will provide a better scaling-solution. It also led many to see Bitcoin as a solid foundation for further development and a storage of value rather than a day-to-day currency, as was initially the case.

5.6. Second-layer solutions

Remember when I said that some issues might only be solved using second-layer or off-chain solutions? Like the fact that blockchains don't scale very well – since there can only be a limited number of transactions per block every few minutes; or that they are slow for over-the-counter payments, since confirmations can take an hour; or that eventually the fees will need to compensate mining activity, meaning they're prone to grow rather than to shrink.

Here is a very brief introduction to second-layer solutions, which, while definitely part of the bigger Bitcoin ecosystem and important in their own right, would require an entire book of their own to explain properly, as they come with advantages and disadvantages away from the blockchain layer.

You can imagine a second-layer solution by picturing the Bitcoin blockchain as a huge tanker, transporting containers.

In this analogy, the second layer acts as trucks transporting smaller loads from the huge tankers to stores and wherever else they're needed.

Second-layer technologies aspire to take over the small and numerous transactions on land – doing the heavy lifting when you want to buy a coffee, or issue a micro-transaction online, or make regular, repeating payments of any kind. They let the blockchain handle only bundled, rarer commits every few hundred or thousand transactions - as a solid basis for trust.

The Lightning Network

Lightning Network is perhaps the most prominent example of second-layer solutions for Bitcoin. Its underlying idea is that not all transactions need to be stored on the blockchain. Imagine your coffee place. You buy a coffee every morning, so you don't want to wait an hour, nor do you want to pay a considerable fee every time. Lightning Network proposes opening something called a "payment channel" between yourself and the coffee place by initially committing money on the blockchain, kind of like opening a deposit. Now, you can transact using this payment channel as often as you want, and it can stay open for as long as it's useful - hours, days, months or decades. The only time you will touch the blockchain ever again will be when closing the channel. Then, you and the coffee place will both record the final status of all occurred channel transactions on the blockchain.

By extending the idea of a payment channel, one can create a huge network of payment channels passing money through each other and only rarely requiring a transaction on the blockchain itself. A very neat analogy used to explain the Lightning Network concept is a counting frame, or abacus.

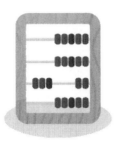

Imagine a payment channel as a single wire on the counting frame depicted in the image above. For simplicity's sake, let's say each bead – the little black and dark grey ones – represents a Satoshi inside the channel. When Alice and Bob create a payment channel between them, Alice deposits five Satoshis inside the channel, by committing them via a multi-signature transaction to the blockchain.

A payment channel between Alice and Bob

A counting frame wire and a payment channel share the following characteristics:

- **Fixed**. Beads cannot be added or removed from a wire. This means Alice and Bob can exchange between them at most the amount of Satoshis that was committed when opening the payment channel. If they want to exchange a larger number of Satoshis, they need to commit more on-chain.

- **Bidirectional**. Like beads on a counting frame wire can be moved left-to-right and right-to-left, Satoshis can be moved from Alice to Bob and vice-versa.

- **Ownership**. When calculating on a counting frame wire, you place the beads either on the left or the right-hand side. Similarly, Satoshis in a payment channel can either belong to Alice or Bob.

If Alice decides to send two Satoshis to Bob, they record a transaction on the Lightning Network (not on the blockchain) and the resulting amounts in the channel look like in the image below. According to the payment channel, Alice now has three Satoshis left and Bob has two.

Alice sends two Satoshis to Bob

Let's imagine Bob has an additional payment channel on the Lightning Network with Charlie, in which he has committed five Satoshis.

Alice, Bob and Charlie form a network

Lightning Network enables the transfer of Satoshis from Alice to Charlie without the two of them having a common payment channel open, as long as there are common nodes along the way – in this case, Bob. When Alice wants to send Charlie two Satoshis, she instead sends two Satoshis to Bob in the Alice-Bob channel. Bob is then required to send two Satoshis to Charlie in the Bob-Charlie channel. This is mandatory since Bob accepted to participate in the transaction. The state of the channels after Alice sent Charlie two Satoshis is depicted below.

Alice can send money to Charlie indirectly, through Bob

Using these simple analogies, it's perhaps easy to see how a network of such channels could be an efficient way to move value globally at a high rate, fast and secure. It is, perhaps, also easy to see how the amounts committed to channels can become a limitation to the network, if intermediary nodes don't have enough capacity to enable payments. It is also easy to see how bad actors could try and commit wrong transaction information to the blockchain, since Lightning Network requires channel participants to be permanently online, otherwise accepting a one-sided commit after a period of a few days.

As mentioned previously, there are a number of advantages and disadvantages, and there exist solutions which go beyond the scope of this book. This is why we will stop here and conclude our observations regarding the technological side of things. Instead, we will move on to something equally as exciting. The economic aspects of Bitcoin.

6. Economic Aspects

6.1. Inflation and Deflation

We've already seen that one of the basic properties of Bitcoin is its scarcity, which makes it somewhat comparable to gold and other raw materials and provides planning security. And since most economic aspects surrounding Bitcoin are in some way related to that scarcity, we need to have a good understanding of what deflation and inflation are before we continue.

Say you finally found the apartment of your dreams, for the modest and oddly convenient – as if to illustrate the point of a book example – round sum of 100.000 euros. It's cosy, with a view of the Mediterranean and the owner, your good friend Atia, promised it to you when she moves out. Now, you're a hard worker and you're smart, so you manage to save 10.000 euros a year. Year after year, you watch your bag of money grow bigger, and after ten years, you're finally ready to buy that apartment! But as you proudly hand over the bag to Atia, she apologetically tells you the apartment now costs 121.899 euros. That's almost 22% more than what you negotiated! Outraged, you leave and secretly blame Atia for her sketchy pricing practices.

So what happened? The good news is, Atia was fair. She put the correct price tag on the apartment. How, you ask? That's the bad news. You just experienced the effect of a 2% yearly inflation rate first-hand and learned a valuable lesson: Inflation means that your money is worth less in the future.

Inflation occurs when the prices of goods and services in an economy go up. To measure it, many governments use a so-called "basket of goods", a collection of goods and services which are representative for the consumer habits in the economy – stuff like TVs, coffee, haircuts, furniture, gas, education costs etc. The recorded monthly or yearly price changes of these goods are then used to calculate the consumer price index (CPI), a measure for inflation. Central banks and governments use the CPI to set the monetary policy – since increasing (or "inflating") the money supply in an economy creates a supply surplus that decreases the value of money (thus prices rise). Decreasing the money supply, on the other hand, increases the value of money (prices fall). This is called a deflation and is represented by a negative CPI.

Western economies usually aim to keep the inflation rate at around 2%. One explanation for this is that too low an inflation would cause consumers to put off spending – hoarding the money instead of reinvesting in the economy. This could cause liquidity issues and drive companies out of business. But some economists, like Milton Friedman, have less flattering explanations, calling inflation "taxation without representation", as it gives governments leeway to literally print money while simultaneously decreasing the value of people's wages and incomes.

6.2. Bitcoin's planned deflation

> *"The main difference between primitive money, such as a proper gold standard, and paper money, such as today's fiat money, is the stretching of the amount."*
>
> *- Schlichter[7]*

After all twenty-one million Bitcoins will have been mined sometime around the year 2140, there will be a permanently fixed quantity on the market - save a few occasional losses. And with rising demand, the fixed quantity inherently leads to an increase in value - we have a planned currency deflation on our hands. But we don't even need to look that far into the future, since even today, despite Bitcoin's still high inflation rate, due to the continuous generation of new Bitcoins, the even higher demand causes the Bitcoin value to climb.

Nobel Prize winner Paul Krugman has criticized Bitcoin's growth and deflationary nature multiple times already: „What we want from a monetary system is not to enrich the people who hoard the money. We want to facilitate transactions to enrich the whole economy, and that does not happen in the case of Bitcoin. [...] If you measure prices in Bitcoin, they will crash. The Bitcoin economy has indeed experienced massive deflation."

---

[7] (Schlichter, 2012)

However, since Bitcoin's inflation rate at the time of this statement was around 10% a year (as we've seen in the chart under chapter 4.3), the statement only describes the effect of Bitcoin's growing adoption, rather than its long-term deflationary effect. Still, deflation can pose an issue, so we will take a look at the long-term effects of deflation in general and in the case of Bitcoin specifically, especially as a consequence of the future stagnating total amount of Bitcoins.

Keynesianism and Neoclassicism

Remember the Keynesians from Chapter 3.2? They were the ones who see the fiscal and monetary intervention of the state as a necessity for the stabilization of economic cycles. When it comes to the subject of deflation, according to Keynesian and neoclassical macro-economic theories, deflation prevents investment and consumption, as the rational behaviour of the individual works against the rational functioning of the economy. Simply put, if your money is worth more tomorrow than it is today, you'll likely keep it safe under the mattress instead of doing something useful with it, like buying a new car or investing in a company. The hoarding of money and the associated lack of consumption lead to recession or depression. The problem with this is the creation of a deflationary spiral by the money itself, which creates incentives for counterproductive behaviour.

England is often cited as an example for having escaped the Great Depression early by abandoning the gold standard and thus "liberating itself of the artificial restrictions of the economy". It's believed that artificial limits which are independent of economic growth make investing in money itself more profitable and thus more rational than investing in manufacturing in the long run. This is how the money supply becomes scarce and the incentives for counter-productive behaviour are further intensified.

Unlike gold, Bitcoin has the advantage of a predictable supply. In addition, ease of divisibility weakens some of the problems of commodity-backed currencies. Nevertheless, predicting how its deflationary nature will play out is perhaps not possible, which is why Bitcoin is still considered by some an interesting economics experiment.

In times of speculation, individual rational behaviour will most probability lead to bubbles that eventually burst. Proof of this statement can be found in the „Hahn problem", named after British economist Frank Hahn. He showed in 1966 that the ability to accumulate multiple assets - like gold and dollars - leads to an instability in economic development if the actors behave rationally and have complete foresight in the short term, for example, in our case, if they would correctly expect the Bitcoin value to rise. Deflation as a result of a bubble can plunge the economy into a depression, so a moderate inflation rate with a positive effect on investment is desirable.

The risk of deflation is only low as long as Bitcoin and other currencies are used in parallel to one another. For example, if the money that has been exchanged into Bitcoins continues to be used for the purchase of goods instead of being hoarded, this could even result in a larger amount of money in circulation and thus create an adverse inflationary tendency.

The Austrian school of economic thought

From the perspective of the Austrian school, on the other hand, inflation is seen as a reason for economic ups and downs. The basic mechanism of interest rates, which signals the time value of money and the available savings, is disturbed by the actions of the central bank. The key interest rate cannot be determined objectively, so misjudgements lead to overproduction or underproduction of goods.

Hülsmann, a supporter of the liberal Austrian economist Ludwig von Mises, describes in his monography "Deflation and Freedom" the social and political benefits of deflation.[8] One of his main arguments is the ability of deflation to eliminate the concentration of revenues resulting from the monopoly position of central banks. Deflation decentralizes financial decisions and makes banks, companies and individuals more prudent and self-reliant in dealing with money than inflation would. Following this line of reasoning, those institutions which are liquidated in a deflation deserve to be liquidated, as they have invested too carelessly.

The creation of money in a free society should therefore be a matter of private initiative, in which market participants invest efforts for mining and coinage and have a competitive character. Also, there should be a completely free association of market participants with the money, e.g. there is no obligation to use it.

This theory is supported by the experience with deflation in the USA, England, and Germany in the late 19th century, which can also be used as a counter-argument to the previously described neoclassical

---

[8] (Hülsmann, 2008)

perspective. This period was characterized by rapid productivity and economic growth despite low deflation. Only the experiences of the 1930s allowed deflation to become synonymous with depression. Some economists claim that only unexpected deflation has negative consequences, whereas the negative money supply shocks towards the end of the 19th century had little effect on production.

Other considerations on deflation

The above-mentioned Keynesian theories are widely used in practice and have experienced a long-lasting economic boom, which is why they are often considered to be valid in economic circles. But existing empirical studies might not be conclusive about Bitcoin. Certain features of Bitcoin may cause a different evolution in deflation than traditional currencies would.

For example, at the moment and for the foreseeable future Bitcoin behaves deflationary only as the Bitcoin economy grows. This, along with divisibility down to a Satoshi which could allow continuous downward adjustments of prices and salaries in a Bitcoin economy, can lead to a slow and steady deflationary rate. There would be little danger of deflation in a Bitcoin economy in which the price behaves predictably.

Then again, the incentive to keep and hoard Bitcoin could be increased by this same high divisibility, because people will assume that it can still be used as a medium of exchange, even in times of severe deflation. The American journalist Surowiecki sees a danger here that Bitcoin could be caught in a deflationary spiral: "With ordinary currencies [...] there is a limit to how far down the spiral can go, as people continue to have to eat, pay their bills and so on. To do this they are forced to use their currency. But these facts do not apply to Bitcoin: people can continue very well without ever spending them, so there is no need to stop hoarding and to begin spending. It's easy to spot a scenario where most Bitcoins are held by people, hoping to sell them to other people." [9]

Following the above deliberations, we can say for sure that we have no idea whether deflation is a lasting effect, or whether a long-term price stabilization can be expected as soon as the dynamic system reaches equilibrium, i.e. has reached a certain price threshold. Don't let anyone claim otherwise.

And finally, as a cherry on top to all these theoretical questions, there is the philosophical and moral question of whether continuous economic growth is even possible and desirable in the long term. Can low deflation be brought forward as a means against artificially generated inflation, even with the risk of losing investment potential? To quote the economist and philosopher Boulding:

> *"Anyone who believes exponential growth can go on forever, is either a madman or an economist."*
>
> *- Kenneth Boulding*

---

[9] (Surowiecki, 2013)

6.3. Scarcity and Stock-To-Flow Ratio

While discussing the way mining works, we talked a bit about halving rewards and the ensuing scarcity. The information and opinions I'm about to present are to be taken with a grain of salt, since Bitcoin is perhaps too young to assess their long-term validity. However, it can be valuable information, so I feel compelled to write about it.

I'm talking about an economic model called Stock-To-Flow, which describes the scarcity of commodities as a function of their *stock*, i.e. the existing amount in circulation, in relation to their *flow*, i.e. the yearly production. According to the model, assets like gold, silver and in our case, Bitcoin, are different than consumable commodities like zinc, copper and even platinum mainly because of their consistently low rate of supply.

For example, there is an existing stockpile of 185 thousand tons of gold available at the moment, but only about three thousand new tons of gold are mined each year, which means gold has an annual supply growth of little over 1.5%. This makes it a scarce asset with a Stock-To-Flow (SF) value of 62, as it takes 62 years to double the existing amount of gold in circulation. Intuitively, the higher the SF value, the scarcer the asset. And gold, as with other commodities, has a certain stability to its SF value dictated by the market: if its demand grows, the resulting price rise triggers a rise in mining and production, which cause the price to fall again and the SF value to return back to its equilibrium value. Silver is similarly trapped at an SF value of approx. 22. The Bitcoin SF value is currently around 25, but since its price has no bearing on the produced quantity – which is slowly falling – its SF value is steadily growing.

Interestingly the SF value can be used to construct a model[10] which highly correlates with the price of the modelled asset, as depicted in the chart below for Bitcoin.

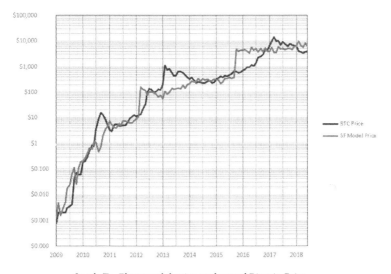

Stock-To-Flow model price and actual Bitcoin Price

Here one can almost identify an occasional rush of the actual Bitcoin price to adjust to the modelled price after each big jump – especially in 2011, 2013 and 2017. The jumps in the modelled price are attributed to the reward halving events, which increase Bitcoin's SF value by decreasing its production.

---

[10] (Modeling Bitcoins value with scarcity, 2018)

When looking at the evolution of the Bitcoin price through the lens of halving-periods, one is perhaps able to read into it - if one is so inclined - a series of repeating patterns, as exemplarily depicted in the chart below. But once again, Bitcoin is not old enough to confirm these patterns, and this is more akin to reading star signs and meridian lines than actual science. And if these patterns prove true, then the more they are confirmed, the more likely they will eventually break down due to traders taking advantage of them.

The price development following each halving

Even if there is some truth in the presented hypothesis, there is no conclusive evidence that scarcity drives value. As an example, the North American indigenous peoples considered gold worthless, and other minerals with higher, more pragmatic uses such as obsidian, flint and slate were preferred and thus more valuable in barter and trade. So even a formidable potential SF ratio can be worth nothing when there is no demand.

6.4. Liquidity and Volatility

Liquidity

The term liquidity describes the availability of assets on a market. In a stock market, underlying assets like physical resources, stocks, bonds etc. are usually only managed and transferred as deposit balances. What this means is, the actual underlying asset doesn't move when you trade it. For example, say you're buying a "Gold ETF", an exchange traded fund.

Simply put, an ETF is a paper (called a derivative contract) that is backed by real physical gold. The physical gold itself doesn't exchange hands. In fact, it doesn't move at all, only the ETF does. The balances are kept in "collective deposit accounts", which are gathered anonymously by a broker and occasionally lent to short sellers, which bet on falling prices. Through the ability of the anonymous lending the broker can ensure a suitable stock of existing orders at any time. He needs this to be able to buy and sell even if an insufficient number of orders would not allow a market order[11].

Similarly, Bitcoin exchanges collect Bitcoins, which are bundled in buy and sell orders. However, unlike with the stock markets, traders rarely keep their Bitcoins on exchanges - a behaviour that can be attributed to the high losses caused by cyber-attacks and bankruptcies of several Bitcoin exchanges in the past and also to the "hodl" phenomenon, which describes individuals hoarding Bitcoin. There is a saying in the Bitcoin world, "not your keys, not your

---

[11] An order which is executed at the price valid on the stock exchange at the time of the order entry.

money", which rightfully encourages users to take their private keys offline and keep them safe.

The resulting low number of available Bitcoins sometimes – though rarely – makes it difficult to fill market orders. Without guaranteed market orders, the Bitcoin market lacks liquidity. A wider range of financial instruments based on Bitcoin (such as futures, options on futures, options on volatility options as well as ETFs, all of which you need not worry about if you've never heard of them) can provide liquidity to the market and facilitate leveraged trading without the need to lend Bitcoins. This is why these instruments are highly anticipated. Another reason is that with such financial instruments, institutional investors are much more likely to enter the market – since they then no longer require technological expertise to trade Bitcoin.

Volatility

A major criticism of Bitcoin's stabilizing ability – or inability – is the fact that governments or banks can't influence its money supply. Some see not only the privilege, but also the obligation and responsibility for state issuers to maintain a stable price level in times of economic expansion and contraction. The more unstable a currency is, the less it can fulfil all three classic monetary functions – functioning as a medium of exchange, unit of account and store of value. Fiat currencies are generally stable when central banks maintain a stable policy and follow monetary policy rules rather than making case-by-case decisions. This is something the literature calls "rules versus discretion", and it cannot be applied to unregulated currencies such as Bitcoin.

However, let me put this critique into perspective by saying it could prove biased. It excludes the possibility of alternative currency rules created by the free-market, which, as far as we can imagine, might a well prove superior. Furthermore, price instability is often referred to as a "childhood disease" of crypto currencies, which is due to disproportionate speculation in the beginnings and which decreases with growing adoption.

In this context, as well as in the question of deflation, different opinions clash, but without sufficient empirical evidence to give credence to the theses. More substantiated evidence is provided by the results of a study by the University of Chicago, which was part of their "Initiative on Global Markets". Hereby thirty-eight leading American economists have been asked about the long-term volatility of Bitcoin. On the statement that „*the value of a Bitcoin is derived solely from the belief that others will use it for trade, which will likely cause its purchasing power to fluctuate over time to a degree that will limit its usefulness*", 61% of the economists said that they agree and 18% that they strongly agree. Only 5% said they disagree, while the remaining respondents had no opinion or were unsure.

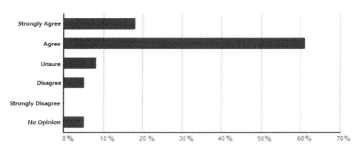

Results of study regarding long-term volatility of Bitcoin[12]

---

[12] (IGM, 2013)

These results provide a picture of the prevailing opinion on Bitcoin. People expect some intrinsic value as the basis for stable purchasing power. When such a value is missing or not immediately visible, as in the case of Bitcoin, its usefulness is questioned.

This is due to the idea that demand for monetary assets is no different than the demand for other assets. But if you agree with this feeling, then consider the value of transaction-facilitating services. Surely the ability to transact quickly, safely and worldwide is itself worth something? Transferring Bitcoin worth a million Euros - or worth any sum, for that matter - from Germany to Australia is a breeze compared to an equivalent amount of gold or any other asset. Or how about the robustness against central control? Historically, dictatorial regimes have made a hobby out of confiscating the gold and other assets of political opponents, a scenario which is unfathomably more difficult and improbable with Bitcoin. Surely, then, these aspects shouldn't be ignored when evaluating its intrinsic value.

Also ask yourself what determines the purchasing power stability of assets such as gold and silver, the value of which derives largely from the same beliefs mentioned in the question above.

Since we can't really foresee the deflationary impact of Bitcoin more accurately than this, we can at most use the stability of gold as an indication of the long-term purchasing power of a scarce asset. We assume that this development can be applied to Bitcoin – at least to some extent - due to the similar characteristics of both assets.

Harmston's study „Gold as a store" finds that despite price volatility, gold has kept its value in terms of purchasing power over time, and over and over has returned to its historical purchasing power parity with other commodities and also to the Consumer Price Index (CPI) basket.

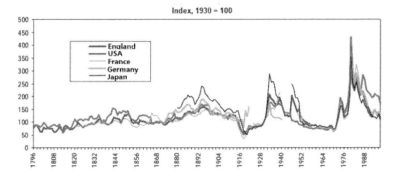

The purchasing power parity of gold to crude oil in the last 40 years and its average fluctuation[13]

6.5. Cryptocurrency exchanges

We have so far discussed multiple economic aspects which are intrinsic to Bitcoin and the way it is technologically constructed. However, influence - and with it risks - to any cryptocurrency is not only derived from within, but also from without, like changes to its ecosystem and to the points of interaction between the digital world and the real world. Negative effects on big cryptocurrency exchanges tend to have a negative impact on the renown, price and acceptance of the asset itself. Three of the biggest risks that Bitcoin exchanges are exposed to are Distributed Denial of Service (DDoS) attacks, the

---

[13] (Harmston, 1998)

defaulting of payment service providers and political regulation, and they have more than once caused issues in the crypto world.

Distributed denial of service

A distributed denial of service (DDoS) attack typically overloads the bandwidth or processing capacity of a system, causing failure of the service or the entire network connection. For example, a web page called up millions of times in a short time would temporarily reject new connections.

For exchanges, such an attack can be achieved using botnets that execute thousands of microtransactions to slow down the exchanges' servers. For example, in 2013 the MtGox exchange had delays of more than 70 minutes and was unreachable for more than 10 hours. This was partly attributed to newly introduced, relative trading fees of 0.5% for a purchase order, which meant that small transactions could be executed in huge quantities - as they cost almost nothing. The goal of such a DDoS attack is to induce panic sales and thus lower the price to the benefit of the attacker.

This is an issue that is easily mitigated by setting a fixed minimum fee on transactions, but which has caused considerable damage in the first years of Bitcoins existence. A system that does not impose any kind of costs on its users invites abuse. Take for example spam emails. The use of a fee (or even better, a proof-of-work mechanism) when the e-mail was invented would have saved tremendous amounts of long-running costs caused by spam mail. Even the internet protocol could have benefited from a minimal fee (or, again, proof-of-work) for client requests, resulting in distributed denial service attacks that would have become unprofitable.

Failure of payment by service providers

Another risk for exchanges, and thus indirectly for Bitcoin, is their dependency on payment service providers. U.S. American services such as PayPal, Dwolla and Liberty Reserve make fiat money transactions in and out of the fiat money accounts of the exchanges and thus into and out of the Bitcoin market. This dependency can affect both liquidity, market share and stock market profitability, should payment options fail.

For example, on May 16, 2013, the exchange platform MtGox lost its account with the payment service provider Dwolla, after the company missed registration as a money service business with the American FinCEN (Financial Crimes Enforcement Network). On May 25, the payment partner of MtGox, Liberty Reserve, was closed for money laundering and on May 27, MtGox lost a third trading partner, the American payment service provider OKPay. In June 2013, MtGox was temporarily forced to stop paying out dollars.

6.6. Further economic considerations related to Bitcoin

In this subchapter we will take a look at a few additional ideas as well as criticism or misconceptions regarding the economics behind Bitcoin. These are Mises' Regression Theorem which attempts to answer the question of whether Bitcoin can be considered real money; the virtually democratic nature of Bitcoin through its network consensus on transaction rules; the Labour Value Theory often attributed to Karl Marx and finally the impossibility to retract transactions.

Mises' Regression Theorem

> *"[...] no good can be employed for the function of a medium of exchange which at the very beginning of its use for this purpose did not have exchange value on account of other employments. And all these statements implied in the regression theorem are enounced apodictically as implied in the apriorism of praxeology. It must happen this way. Nobody can ever succeed in constructing a hypothetical case in which things were to occur in a different way."*
>
> *- Ludwig von Mises*

You're probably thinking, "apodictically implied in the aprio-what of prax-whom?". The purpose of the Regression Theorem was to solve the apparent paradox of money, in which money acquires value as a means of exchange, despite it being valuable only because it serves as a means of exchange. Put more simply, why should I believe that an asset can have value and function as a means of exchange, when that exact same belief is the only thing that gives it value in the first place? What a confusing, circular dilemma.

Mises solved this apparent circularity by pointing to the critical, missing temporal component.[14] He pointed out that at some point in the past, money must have been related to a commodity. This is what the term "regression" refers to – any currency must be somehow historically connected to an initial thing that had its own bartering value, which it achieved through its direct utility or consummation.

According to Rothbard, all future price estimates of value as a medium of exchange arise from that original value. As he explains in *Man, Economy, and State*, „the economic analysis of monetary values is [...] not circular. If today's prices are based on the marginal utility of today's money, then the latter is based on yesterday's prices."[15]

And because Bitcoin is not directly related to a direct utility product, critics argue that Bitcoin does not meet the requirements of a medium of exchange. They rely on Ludwig von Mises' Regression Theorem to argue that Bitcoin cannot be real money and eventually will lose its value.

It is easy to refute this argumentation if the Regression Theorem is understood as an apodictic statement, e.g. a self-evidently, absolutely

---

[14] (Von Mises, 1998)
[15] (Rothbard, 2009)

certain statement, as intended by Mises. This means that Bitcoin, which in fact already is a means of exchange, must necessarily have had a direct utility value or direct exchange value in the past. A failure to find such a value is therefore due only to insufficient investigation on our side and makes no statement about the former existence of the value.

The attempt to determine the former direct benefit of Bitcoin, which undoubtedly exists according to the Regression Theorem, is therefore an interesting thought experiment but not imperative for the suitability of Bitcoin as a medium of exchange and we could therefore refrain from addressing this further beyond this point.

But just in case this topic caught your attention, here are a few more ideas worth considering. On the one hand, the initial translation of Bitcoin into fiat money via exchanges could explain the emergence of a currency based on existing currencies. For example, this explains the transition of the German mark and the French franc to the euro, without the euro ever having had a direct utility. This consideration implies that an exchange already taking place between two currencies, such as the Bitcoin and the U.S. dollar, complies with the monetary regression, as this may indirectly relate to barter goods.

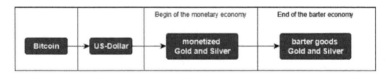

Possible monetary regression of Bitcoin

Another interesting point is the consideration of whether features such as novelty, the ability to awake interest, competitive advantage, etc., may or may not initially be valued, so that the value of gold or

Bitcoin need not go back further than their useful existence as a medium of exchange.

Consensus on the rules

A common misconception is the assumption that the rules of the Bitcoin system were initially fixed and are not changeable. The original rules, ranging from the maximum number of Bitcoins, to the amount and frequency of change of the rewards and to the smallest currency unit, were determined by Satoshi Nakamoto. However, these could be changed if enough network participants agreed to accept other rules. The rules are not self-executing, so to be applied, more than half of all network nodes must ignore the existence of non-compliant transactions while accepting and verifying valid transactions. Remember the byzantine generals? Accepting only certain rules happens automatically during mining by the deployed clients implementing the agreed rules. The clients check the compliance of incoming transactions for these rules.

The ability to change rules is not just theoretical, as the rules were already changed in the past. So it's easy to see how these rule-changing possibilities raise several questions that require closer consideration:

- How does a rule change?

- Who has the authority to make such changes? To what extent is the decision truly democratic despite the mining pool oligopoly and the existence of computationally strong companies?

- To what extent does this pose a risk to the stability and integrity of the currency?

- To what extent does this allow the currency to adapt to changing market conditions?

In this context, a source of uncertainty is the ability of a few mining pools and companies to hold a large part of the total network capacity. As a result, they would have disproportionate voting rights in any rule change.

Luckily, the individual mining pools are bound to their user base and risk losing their trust if breeches occur, which makes voting against the interests of the user base less likely. On the other hand, some mining companies have no restrictions in this regard and can freely decide which transactions to accept.

We could say that, as in a democracy, consensus rules work great only as long as enough people participate.

Labour Theory of Value

The Labour Theory of Value is an approach of classical economics to determine the value of a commodity by the power used to produce it. The best-known advocates of the Labour Theory of Value were Adam Smith, David Ricardo and Karl Marx. It contradicts the Theory of Marginal Utility and is therefore often described by modern economists as invalid.

For example, in the case of Bitcoin, an attempt is made to pin the real value of Bitcoin to the cost of electricity and computational power used to generate it. However, the value of a commodity might derive from its utility, so the cost of Bitcoin production depends on the Bitcoin value, not the other way around. Meaning a higher Bitcoin price leads to a higher incentive for mining, which entails more computing power and thus an increased degree of difficulty. This increases the production costs in proportion to the Bitcoin value. The opposite happens when the value of Bitcoin decreases, so that the production costs are always adjusted to the Bitcoin value.

No retraction of transactions

While this specific characteristic of Bitcoin - the irreversible nature of transactions - provides security for the seller, transactions without middlemen or other service providers may prove risky for the buyer. The buyer has no opportunity to cancel his payment if the goods are not delivered. In the opinion of the U.S. Department of the Treasury, this feature makes Bitcoin attractive for criminal use and problematic for legitimate payment functions.

In the Bitcoin economy, escrow services have established themselves as a solution to this lack of trust, forcing payment to be held until the contractual agreements have been fulfilled. Thus, this problem is negligible. From a technological point of view, Bitcoin also allows a 2-by-3 signing mechanism for transactions, in which a trustee must intervene only if there is disagreement between the contracting parties. This mechanism exists but is hardly used.

7. Suitability as Money

7.1. Economical suitability as money

> *"I have always been in favour of the abolition of the Federal Reserve and its replacement with a machine program that ensures a steady growth within the money supply."*
>
> *- Milton Friedman*

Barter goods that are generally accepted as a medium of exchange, like gold and silver, are defined as money. But while the concept of „medium of exchange" can be described in great detail, the definition of „money" is much less precise. The moment in time in which a medium of exchange is generally recognized as money can't be strictly determined. This is especially true if the currency's acceptance can't be measured in a geographical or socioeconomic setting - for example, a country in which the majority of the population regularly uses the currency. Bitcoin is being increasingly used in many countries but can't be pinned to a single one. This raises the question of whether Bitcoin should be seen in a global context or in the context of individual local economies. Also, the question arises as to whether the recognition Bitcoin obtained as a currency in a local context gives it the money role in general.

Rothbard describes the assessment of when a medium of exchange is money, as a research task for historians. The future and general suitability as money cannot be estimated from current acceptance statistics, but rather from the challenges elaborated earlier.

For the future increase in socioeconomic acceptance, for example, the liquidity and thus the convertibility into existing fiat currencies poses a major problem. Providers of goods rely on being able to exchange to other currencies as long as only a limited number of goods can be bought with Bitcoins. Traders are therefore restricted by the prevailing exchange oligopoly which in turn impairs the acceptance by the demand-side. This makes centralized cash flow a big economic challenge for Bitcoin.

As explained earlier the fixed money supply is also seen as very controversial. A statement about the advantages and disadvantages of deflation can only be made with difficulty due to a lack of empirical observations. There are both opponents and proponents of a fixed amount. McMillan, former Stanford economist of the U.S. Federal Trade Commission, assumes that „the known and limited amount of Bitcoin" represents the downfall of the system „as a viable currency substitute".

Bishop, the editor of The Economist, however, believes that „the algorithmic approach used by Bitcoin to control the money supply […] could do much to create a solid store of value." Gordon, a professor at the Columbia Graduate School of Business, also sees an advantage in the deflationary nature of the system.

Doubts about Bitcoin's suitability as a currency due to deflation have already led to the creation of alternative cryptocurrencies, which try to solve this problem using a permanently inflationary policy. For example, Stellar Lumens has a built-in yearly inflation rate of 1%. Freecoin copied most of the Bitcoin code but removed the fixed money supply and instead limited the difficulty of creating new currency units. This created an inflationary currency whose inflation rate is dictated by market forces. But as of now, we really can't say whether or not deflation will turn into a real issue.

7.2. Technological suitability as money

Starting with the solution offered earlier for finding consensus in a distributed network, the question arises as to whether this is practicable. The six-hour fork of the Bitcoin blockchain in March 2013 showed the potential risk for traders and the necessity to close deals quickly. On-demand purchasing is increasingly suffering as the currency becomes more and more popular and transaction processing takes longer.

Even if Bitcoin does not solve the problem of the byzantine generals to the satisfaction of theoretical mathematicians, the solution offered to reach consensus is practical, at least from a technological point of view. Irrevocable decisions by market participants, such as the switch to a new mining client or the acceptance of large amounts of payment within a single hour, are to be made in accordance with the personal risk analysis and are not considered technical failure of the protocol in this sense. Trust in the security mechanisms of a crypto currency is comparable with the trust in today's online banking, with the difference that Bitcoin is most likely superior in terms of safety.

Although many businesses are already accepting unconfirmed transactions, it can be argued that this technological aspect is a barrier to the long-term acceptance of Bitcoin as a currency, as it creates fraud incentives. Even though the risk of fraud can be kept low when there is a bare minimum exchange of information between the seller and the buyer, e.g. if the payments are not completely anonymous, this requires a minimum of trust between the parties. The trust problem has already been mentioned by Satoshi Nakamoto in his whitepaper, where he recognized that such bundles of confidence create incentives for the abuse of power. Only potential second-layer solutions will prove if there is a viable solution to this problem.

Another danger we have identified earlier is a potential politically motivated act against Bitcoin to occupy more than 50% of the network. Because Bitcoin is used in illicit activity - from payments for pirated content on Usenet, to ransomware attacks demanding payment in crypto, to money laundering and the dark web trade of drugs and weapons - and also because states are inclined to want to maintain control of the flow of capital, there is some incentive for states to curb the adoption of Bitcoin as currency, for example through technological means. However, even with the necessary funds, potential attackers most likely have no way to catch up with the network's hash rates. This is perhaps most visible when considering the time which would be required for planning and acquiring the necessary technologies such as ASICs. The continuous chip production of the free market participants increases the hash rate at a speed with which institutions could not keep up.

Another concern for the acceptance of Bitcoin as a currency is the traceability of transactions, which is a technological reality on a blockchain. However, comparing the blockchain transparency to the current accounting requirements of companies to publish liquidity data in national newspapers, this could be considered non-critical, at least for legal entities like companies and institutions. The traceability of the transactions of private individuals, on the other hand, is a paradigm shift that challenges legal limits. It has, however, little bearing on the technological usefulness of Bitcoin as a currency.

Finally, the incentive problem persists, possibly leading pools to withhold high-cost transactions against the interest of the network.

The attempts could be short-lived, as they would result in the churning out of customers of dishonest pools as well as exchanges.

So, at least from a technological standpoint, the bottle neck which is the amount of transactions that can be verified seems to be the real issue challenging Bitcoin's role as currency, while remaining inconsequential for its role as a store of value.

7.3. Political suitability as money

"Virtual currencies are no riskier than any other electronic payment system."

- *Emery S. Kobor (Strategic Policy TFFC)*

According to a study by "Javelin Strategy & Research", the share of physical money in global transactions is continuously decreasing. Some governments, like the Swedish and the Nigerian, have the abolition of physical money as their goal. And in Europe alone, the volume of non-physical transactions increases 7% per year despite economic stagnation. This shows a general willingness of governments to embrace digital money. However, Bitcoin retains challenges that governments seek to eliminate by replacing physical money with government-backed digital money. According to the FBI, this includes the role of Bitcoin as a means for criminals to move, steal and launder funds or make donations to unlawful organizations. Increasing demand could also make Bitcoin attractive outside of the internet, making it more difficult for law enforcement to identify suspicious behaviour. In short, Bitcoin only allows prosecuting malicious behaviour if the exchange to fiat currencies takes place via regulated exchanges which verify and save personal data. This way, the traceability of transactions through the blockchain can become a law enforcement tool.

Another critique from the political perspective is the fact that it's not possible to implement a tax system within Bitcoin. Until now, transactions can only be levied in the form of value-added tax at the point of sale, which circumvents the socio-political aspects of revenue taxes and other forms of taxation and creates potential for the emergence of black markets. This could lead to politically motivated attempts to legally ban the currency should fiat currencies be threatened.

In addition, the economic system relies on trust in the continued existence of the system, its preservation of value and its stability. Authorities issue guarantees to assure these aspects. In the case of Bitcoin, the continued existence of the currency is guaranteed by its decentralized nature. But this alone does not guarantee that it will preserve its value. It's therefore unlikely that technological trust can quickly replace political confidence - simply because technological trust involves a basic understanding of technology which most people don't possess. Only in time we'll see if the longevity of Bitcoin suffices to build that trust, even without understanding the technology behind it.

8. The development of the Bitcoin Economy

8.1. Key events and price development

Listed below are selected key events that played a role in Bitcoin's performance. Due to the huge number of potentially important events, the list is incomplete and doesn't cover all relevant or influencing events. So please understood it purely as an author's overview of what happened up to the moment of writing. For the sake of clarity, the chronological development is divided into a few subsections.

2008 – 2010

- Satoshi Nakamoto publishes the Bitcoin white paper on 31 October 2008.

- On January 03, 2009, the genesis block is created.

- Five days later, on January 08, 2009, the Bitcoin code is released.

- On January 12, 2009, the first Bitcoin transaction between two people occurs, when Satoshi Nakamoto sends 50 BTC to the cryptographer Hal Finney. This is the only transaction known to have been sent by the creator of Bitcoin.

- In May 2010, the first transaction for real world goods takes place with the purchase of two pizzas for 10.000 Bitcoins by the programmer Laszlo Hanyecz. The price for the pizzas was negotiated on the "Bitcointalk" forum.

- In October 2010, the first escrow exchange takes place on the Bitcoin forum.

- On November 6, 2010, the Bitcoin economy exceeds USD 1 million.

- On December 16, 2010, the first Bitcoin pool, Slush, verifies its first block.

- On February 9, 2011, the Bitcoin price hits U.S. dollar parity. This event prompts several IT journals (Slashdot, Hacker News) to report about Bitcoin.

- In March and April 2011 exchanges are opened to new currencies (British Pound Sterling, Brazilian Reals, Polish Zloty) and TIME Magazine publishes an article about Bitcoin.

As a result of these and other events, the Bitcoin value on the exchange MtGox, which was the leading Bitcoin exchange at the time, hits an all-time high of $31.91 with a market capitalization of approximately $206 million on June 8, 2011.

2011 - 2012

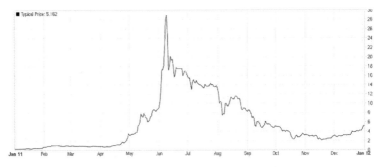

Price Development of Bitcoin in 2011[16]

Beginning in June and ending in late 2011, the Bitcoin value has its first major slump after reaching a low of $1.99 on November 17, 2011. A conceivable explanation for the negative development is a series of thefts.

- MtGox, on June 19, 2011 has 60.000 user accounts stolen from its database. An unidentified hacker is able to log in to MtGox as an administrator and set sales orders of hundreds of thousands of Bitcoins into the system. As a result, MtGox freezes trading for seven days and reverses the sales orders.

- On June 20, 2011, the EFF (Electronic Frontier Foundation) announces they no longer accept Bitcoin payments for donations.

---

[16] (Bitcoincharts, 2019)

The opinions of the media on Bitcoin are largely negative after these events. The online magazine „Wired" publishes an article titled „The Rise and Fall of Bitcoin", which entails statements that the digital currency has become unusable.

Contrary to expectations, the currency stabilizes over the following weeks. After the low is reached at just below two U.S. dollars, the price recovers quickly up to a high of $7.22, when the currency is featured in an American legal drama with over 10 million viewers. A price correction brings the price down again to about $5, where it stays relatively stable for more than three months.

Only in May 2012 does the price begin to rise again.

Price Development of Bitcoin in 2012[17]

By the end of August 2012, the price has risen to $15.4, fuelled by new media attention and an artificial boom brought about by fraudulent

---

[17] (Bitcoincharts, 2019)

Bitcoin Ponzi schemes. When the Ponzi schemes are exposed and collapse, the price oscillates between ten and fourteen U.S. dollars for several months.

In 2013, the price rises almost exponentially, from under $20 in early February to over $220 in April. A popular explanation of this increase is the financial crisis in Cyprus. Another explanation by economists are the new investors stepping into the Bitcoin market after the Financial Crimes Enforcement Network (FinCEN) makes clear statements in March 2013 regarding the legal status and legal obligations of decentralized currencies. Added to this is the growing number of companies that begin accepting Bitcoin as a payment method together with the interest of the media and the general public. In October 2012, Bitcoin is listed as a currency in Capgemini's eighth annual World Payments Report.

2013 - 2014

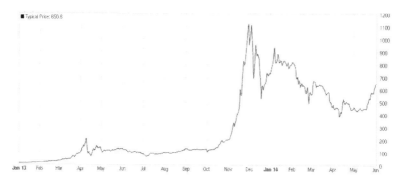

Price Development of Bitcoin in 2013-2014[18]

In April 2013, the payment processor „BitInstant" and the largest exchange platform MtGox experience strong delays of up to a half an hour on sales orders due to high demand overburdening their capacity. As shown above the Bitcoin price falls from $266 to $76 on April 10th, 2013 and returns to $160 within six hours.

Following the collapse in April 2013, Bitcoin experiences a long period of high volatility coupled with low liquidity. Silk Road, which was one of the largest black markets using Bitcoin as a method of payment, is forced by DDoS attacks to shut down its services at short notice. This prompts dealers to leave the Bitcoin market.

- It is publicly announced that sensitive data has been stored in the blockchain, including 2.5 MB of Wikileaks U.S. diplomatic

---

[18] (Bitcoincharts, 2019)

cable leaks as well as links to child pornography. Once this data is committed to the blockchain it cannot be removed, meaning every owner of the blockchain has access to it. This results in damage to the image of the blockchain.

- Two of the largest Bitcoin companies at the time, CoinLab and MtGox, have litigation over their contractual arrangements. U.S. Homeland Security blocks an account of the MtGox Mutuum Sigillum LLC subsidiary after the exchange fails to register as a money transmitter.

2015 - 2016

Price Development of Bitcoin in 2015-2016[19]

- In this period, multiple exchanges suffer major hacks and come under stronger regulatory scrutiny. The exchange Bitstamp suffers a major hack and loses 19.000 BTC, or about $5M at the

---

[19] (Bitcoincharts, 2019)

time and Bitfinex loses over $60M. The IRS demands user information from all exchanges.

- Despite this, big players garner strong interest in Bitcoin. Wallet and exchange provider Coinbase raises funding from investors including the New York Stock Exchange (NYSE), Fortune 500 financial services group USAA, Spanish megabank BBVA and Japanese telcom giant DoCoMo. Later, multiple big institutions like the global stock market Nasdaq, the Bank of England and IBM, begin experimenting with blockchain technologies. Multiple blockchain projects like Ethereum and ZCash take off.

- In a series of clashes between powerful names in the crypto space and the law, the founder of the online dark market Silk Road, Ross Ulbricht, is sentenced to life in prison. MtGox founder Mark Karpeles is charged with embezzlement for tampering with user balances for personal gain, leading to the shattering collapse of MtGox.

- A controversial attempt is made to resolve the scalability issue, when Bitcoin developer Mike Hearn forks Bitcoin XT to allow an increase of the block size limit.

2017 - 2018

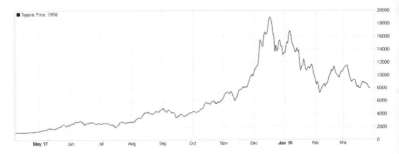

Price Development of Bitcoin in 2017-2018[20]

- More and more coins and tokens begin appearing in the crypto space. "ICOs" (Initial Coin Offerings) take flight and with them a new exchange, Binance.

- In 2017, China, which has become the biggest force in the crypto world with 8 out of 10 exchange transactions being in yuan, announces it would ban cryptocurrency exchanges. Except for a short period of panic, this has no tangible, lasting effects.

- Bitcoin Cash and Bitcoin Gold fork from the Bitcoin blockchain. Bitcoin Cash claims to solve the scalability issue by increasing the block size and causes a big ideological rift in the Bitcoin community.

---

[20] (Bitcoincharts, 2019)

- In January 2018, after an entire year of solid growth, the Bitcoin price falls from a high of nearly $20.000 down to below $8.000 in March and stays well below or around $10.000 until the end of 2019.

8.2. Adoption and use cases

Not only the Bitcoin price, but also the socio-economic acceptance of Bitcoin has increased over the years. The number of merchants accepting Bitcoin as a means of payment has risen, with Bitcoin establishing itself in many industries outside of the online businesses where it was initially pursued. This development combined with its increase in market value in recent years indicate a general increase in popular confidence in Bitcoin.

Among the major economic milestones of Bitcoin is its acceptance as a means of payment by a number of high-ranking companies and the repeated fall-back onto Bitcoin as a crisis currency for countries experiencing hyperinflation and recessions.

Crisis currency

Ever since Bitcoin stepped onto the public scene, economic and political circumstances have led investors in multiple countries to use Bitcoin as a crisis currency and as a preferred long-term investment over their national currencies. In such cases, decreasing popular trust in the national fiat currencies and in the banking-system has strengthened the value, the volume of exchange and the acceptance of Bitcoin as a medium of exchange and store of value.

According to reports from several journals, in 2013 the sudden increase in Bitcoin demand is strongly linked to the banking crisis in Cyprus at the time. Following the announcement by Cypriot banks that a special fee should be levied on all bank balances, the Bitcoin price rose by over 20 percent within a week. The banks in Cyprus went on to freeze bank accounts, which only caused more uncertainty and lead investors in neighbouring countries to switch to alternative investments.

During the same period, demand for Bitcoin skyrocketed in Argentina, at a time when the high inflation rate was kept a secret by the Argentinian government, but which is estimated by the CIA to have hovered above 25%. The interest of the Argentinian population in Bitcoin grew as the people attempted to circumvent government bans on credit cards for international transactions as well as attempts of the government to forbid exchanging Argentine pesos for gold or dollars.

Very similarly, in 2019 Venezuela has seen its currency rendered valueless after one of the worst hyperinflations since WWII. A cup of coffee suddenly cost 2.800 bolivars, up from only 0,75 bolivars twelve months before, despite a 2018 devaluation that had already knocked five zeroes off the currency. More than three million Venezuelans left the country, as essential goods such as toilet paper and medicine become unaffordable, and many turned to digital assets such as Bitcoin as a vehicle to conserve value. Even the government launched its own crypto-currency, the Petro, supposedly backed by oil, to provide a solution to the economic crisis.

Industry sectors

Beyond the internet business, Bitcoin finds acceptance in a variety of industries. Products traded in Bitcoin include precious metals, clothing, games and accessories, office supplies, fitness equipment, artwork, electronic devices, automobiles, real estate, digital downloads, social networks and dating sites, and many more.

Drivers of this growth are venture capital companies as well as individual company investors who finance Bitcoin start-ups.

Financial service providers

In finance, there is an interest in using blockchain technology to reduce transaction fees and enable transactions that are not centrally reviewed and controlled. Bitcoin exchanges partnering with payment services may function as a bank, create payment accounts and assign them to IBANs and credit cards. This can allow credit cards and state-insured accounts for cryptocurrencies.

Other companies can open current accounts with partner banks so that the trading and processing of transactions can run through these banks.

Donations

Bitcoin has increasingly been accepted by charities such as the Electronic Frontier Foundation, the Child Play's Charity, the Bitfilm Festival, and the German Federal Association for the Environment and Nature Conservation e.V., one of Germany's largest non-profit making organizations.

The Electronic Frontier Foundation temporarily discontinued Bitcoin donations in the past due to the unresolved legal situation and concerns about certain features of Bitcoin. The main concerns were the lack of sufficient anonymity, the amount of computing power required and the deflationary nature that rewards hoarding Bitcoin. Acceptance of Bitcoin donations were reinstated.

Even organisations that cannot or hardly accept donations through traditional payment methods are increasingly turning to Bitcoin. One example is Wikileaks, which has been accepting Bitcoin donations since 2011 in order to circumvent account defaults by payment service providers such as PayPal.

Alternative uses of a blockchain

In addition to being used in the industries described, blockchain technology allows a wide range of applications beyond functioning as a purely financial instrument.

- A blockchain can be used as a triple-entry accounting system. This can increase security and reduce the cost of accounting, as well as provide greater transparency for stronger corporate governance.

- In the blockchain, hashes of documents, e.g. contracts, can be integrated as a permanent notarial authentication. Thus, the creation time and the immutability of the document can be shown.

- A 2-of-3 signatures scheme for issuing money makes distributed property and trustee services possible, in which the trustee must intervene only in the event of disagreement.

- A marked Bitcoin can be used to represent something else, like a share. It can be issued, sold, transferred and traded without a bank or broker. Such markings are already being used via so-called „coloured coins".

- The blockchain and concept of proof-of-work can be changed to send messages instead of financial transactions. This enables encrypted, anonymous and distributed e-mail systems.

- The blockchain and the concept of proof-of-work can be changed to include Domain Name System (DNS) information instead of financial transactions. The DNS is basically the phonebook of the internet, which translates IP addresses to website domains like www.reddit.com. Attempts to establish a decentralized DNS system using the blockchain have already been made, but so far has not been able to establish itself, due to being misused as data storage.

- Bitcoin allows local storage of value in software. Applications can thus have monetary value and send it directly without the detour via a bank or a central office. The social media site Reddit, for example, allows its users to give each other small Bitcoin donations as a "tip" for particularly outstanding contributions.

9. Closing Words

In the money business, trust is everything. Many people bring up the value of gold and try to pin it on practical uses in jewelry, dental crowns or even microchips and other applications, many of which didn't even exist a - historically - short while ago. But this is a fallacy of causation. Gold isn't valuable because it's used in jewelry - it's used in jewelry because it's valuable. Gold is valuable because during thousands of years of being used as store of value and medium of exchange, it could never be forged or reconstructed chemically - though many alchemists have tried throughout history. Its resilience has cultivated trust in its long-term ability to maintain its money properties, and trust cultivates value.

Bitcoin is the first scarce digital good and the oldest cryptocurrency in existence, therefore it garners the most trust in its safety – it has never been successfully hacked. And every passing day, it proves more and more that it's robust and becomes more trustworthy.

Additionally, because of its longevity and first mover advantage in the crypto space, Bitcoin also has the biggest developer communities, the highest number of ongoing improvement developments and the strongest brand recognition worldwide. It still often represents the only trading pair that can be used to buy or sell more exotic crypto coins on exchanges and is the only cryptocurrency which institutions seem genuinely interested in investing in, most likely thanks to its brand recognition and trust.

And while blockchains themselves don't seem to be the best solution for over-the-counter trades and coffee purchases - tasks which will hopefully be successfully handled through second-layer and other solutions - they seem to have discovered their place as assets for reliably storing value.

In the author's opinion, Bitcoin is a worthy competitor to gold - except it's simpler and cheaper to produce - all costs considered; simpler and cheaper to store - all costs considered; simpler and cheaper to transfer worldwide - all costs considered. Bitcoin is a global store of value and its existence as such is guaranteed so long as no technological loopholes are found. Its value is most likely going to continue its path of growth while Bitcoin finds its permanent place among other means of value storage - and will chip a little bit out of the gold market, a little bit out of the fiat market, a little bit out of the remittance and SWIFT markets and a little bit out of the black markets.

Some of its value will come out of countries hit by hyperinflation and some out of countries hit by trade bans. Some out of offshore banking and some out of illicit uses. Some will come from people who need to escape their national currency, some from stock traders hedging against other markets and some from anyone hedging against recession and economic collapse.

Acknowledgements

For the limitless support, ideas, food for thought, interesting discussions, reviews and guidance, I'd like to thank Ciprian Ungureanu, Raluca Aldea, Mateusz Krakowiak, Martin Spitzenpfeil, Mark Keinhörster, Sebastian Kurz, Doina Prusan, Frank Hecker, Alin Dobre, the Bitcointalk Community and last but not least the /r/Bitcoin and /r/Bitcoinmarkets subreddits.

References

Becker, J., Breuker, D., Heide, T., Holler, J., Rauer, P. H., & Böhme, R. (2012). Can We Afford Integrity by Proof-of-Work. Scenarios Inspired by the Bitcoin Currency. *Workshop on the Economics of Information Security WEIS 2012.* Berlin.

Bitcoin Inflation. (2015). Von http://www.mattwhitlock.com/Bitcoin%20Inflation%20logarithmic.pdf abgerufen

Bitcoin Stats. (2013). Von http://www.bitcoinstats.org abgerufen

Bitcoincharts. (2019). Von http://bitcoincharts.com/ abgerufen

Blockchain Info Pools. (03 2018). Von http://blockchain.info/pools abgerufen

Friedman, M. (1960). *A program for monetary stability.* Fordham University Press.

Harmston, S. (1998). *Gold as a store of value.*

Hash Rate. (2019). Von Blockchain.info: http://blockchain.info/charts/hash-rate abgerufen

Hülsmann, J. (2008). *Deflation and Liberty.*

Icons made by Freepik, & Payungkead. (2019). *Flaticon.* Von www.flaticon.com abgerufen

IGM. (2013). *Initiative on Global Markets. Bitcoin.*

Lamport, L., Shostak, R., & Pease, M. (1982). *The Byzantine Generals Problem.* .

Modeling Bitcoins value with scarcity. (2018). Von https://medium.com/@100trillionUSD/modeling-bitcoins-value-with-scarcity-91fa0fc03e25 abgerufen

Rothbard, M. N. (2009). *Man, Economy, and State. A Treatise on Economic Principles. With Power and Market. Government and Economy.*

Schlichter, D. (2012). *What gives money value, and is fractional-reserve banking.* Von https://www.cobdencentre.org/2012/03/what-gives-money-value-and-is-fractional-reserve-banking-fraud/ abgerufen

Steadman, I. (2013). Bitcoin interest spikes in Spain as Cyprus financial crisis grows.

Surowiecki, J. (2013). *Cryptocurrency. The bitcoin, a virtual medium of exchange, could be a real alternative to government-issued money. But only if it survives hoarding by speculators.* Von http://www.technologyreview.com/review/425142/cryptocurrency abgerufen

Von Mises, L. (1998). *Human Action. A Treatise on Economics.*

Made in the USA
Coppell, TX
22 October 2020